MISSION TO SATURN

MISSION TO SATURN

The Story of a Debate about Science and God

Arnold Benz and Samuel Vollenweider

A Crossroad Book
The Crossroad Publishing Company
New York

The Crossroad Publishing Company www.crossroadpublishing.com

© 2022 by Arnold Benz and Samuel Vollenweider

This book is similar to an earlier book by the same authors in German (*Würfelt Gott?* Patmos Verlag, 2000). It is not a translation but a thoroughly revised and partially new book.

Crossroad, Herder & Herder, and the crossed C logo/colophon are registered trademarks of The Crossroad Publishing Company.

All rights reserved. No part of this book may be copied, scanned, reproduced in any way, or stored in a retrieval system, or transmitted, in any form or by any means, electronic, mechanical, photocopying, recording, or otherwise, without the written permission of The Crossroad Publishing Company. To inquire about permissions, please email rights@crossroadpublishing.com.

In continuation of our 200-year tradition of independent publishing, The Crossroad Publishing Company proudly offers a variety of books with strong, original voices and diverse perspectives. The viewpoints expressed in our books are not necessarily those of The Crossroad Publishing Company, any of its imprints or of its employees, executives, owners. Although the author and publisher have made every effort to ensure that the information in this book was correct at press time, the author and publisher do not assume and hereby disclaim any liability to any party for any loss, damage, or disruption caused by errors or omissions, whether such errors or omissions result from negligence, accident, or any other cause. No claims are made or responsibility assumed for any health or other benefits.

Cover design by George Foster
Interior design by Tim Holtz

Library of Congress Cataloging-in-Publication Data available from the Library of Congress.

Tradepaper ISBN   9780824550554
epub ISBN   9780824550561
mobi ISBN   9780824550578

Books published by The Crossroad Publishing Company may be purchased at special quantity discount rates for classes and institutional use. For information, please email sales@crossroadpublishing.com.

For Betty and Ruth

Contents

Preface	xi
Crisis on the Elpís	1
An Unusual Crew of Astronauts	4
The Records of the Conversation	7
Creation Today: An Easter Debate about a Hurricane on Saturn (April 2, 2051)	11
An Eighth Day of Creation	19
In Saturn's Sphere: From Thomas's Diary (April 2)	25
From the Old Cosmos to the Modern Universe	35
A Hymn on the Cosmic Christ: From Thomas's Diary (April 4)	37
Creation in Different Worldviews	43
From Darkness to Light: A Postcard from Saturn for the Cyber School	48
Different Perceptions in Different Perspectives	51
A Monster Moves: A Postcard from Saturn for the Cyber School	57
The Language of Images	61
Astrology—a Contentious Point: From Thomas's Diary (April 13)	68
Quantum Mechanics and Reality	71
Quanta Knock at the Door: A Postcard from Saturn for the Cyber School	76
And Endlessly Surges the Quantum Ocean	77
Dance with the Floods: From Thomas's Diary (April 17)	83
Creation without Interruption	85
Why So Uncertain?	93
Physics or Mysticism? From Sheldon's Notebook (April 18)	95

Our Mother, Thou Who Art in the Depth: From Thomas's Diary (April 19)	99
Birth of a Titan: A Postcard from Saturn for the Cyber School	101
In the Beginning Was the Vacuum	105
At the River of Time	115
Cronus Alias Chronos: From Thomas's Diary (April 23)	120
The Birth of Time	123
Does God Play Dice?	131
Between Space and Time: From Thomas's Diary (April 29)	138
A Model of Reality: A Postcard from Saturn for the Cyber School	142
Models—Constructs or Proxies for the Truth?	145
In the Midst of Chaos: A Postcard from Saturn for the Cyber School	148
Chaotic Prospects	151
The Beautiful World of Saturn's Moons: A Postcard from Saturn for the Cyber School	156
In the Shadow of Catastrophes	159
Groping in the Depth: From Thomas's Diary (May 13)	166
Landing on Titan: From Sheldon's Diary (May 14)	166
Extraterrestrial Intelligences and Their Religion	169
Pentecost Is Near: From Thomas's Diary (May 14)	174
Life—the Most Beautiful Child of the Universe	179
A Vision of Titan: From Thomas's Diary (May 15)	183
"And God Saw That It Was Good"? Creation Is Evaluated	185
God—Mighty King or Poor Wanderer?	191
The Universe's Hospitable Nature	197
Homesick for Paradise: A Postcard from Saturn for the Cyber School	203
Götterdämmerung of Humanity?	205
Titan's Shadow over Humankind: From Thomas's Diary (May 17–18)	210

At the Dawn of the Last Days—an Apocalyptic Dream: From Sheldon's Diary (May 19)	213
In the Chat Room of the Journalists (June 19)	215
Logbook Entries from May 19–20	219
One Hour Later in the Chat Room of the Journalists	223
Apocalypse between Physics and Theology	225
Descending into Hell: From Thomas's Diary (May 19–20)	237

The Outcome of the Saturn Mission, or a Space Odyssey 2051   243

Abbreviations and Acronyms	253
Notes	255
Index	269

Preface

Dear Reader,

You hold an unusual book in your hands. It tells the imaginary story of a mission in the year 2051 to planet Saturn and its illustrious host of moons. Two astronauts—one a physicist and the other a theologian—are stranded in their capsule and inspired by the gaseous hurricanes and other phenomena they observe around them in the cosmos to debate their views about God and the universe.

Initially we authors—an astrophysicist (Arnold Benz) and a New Testament theologian (Samuel Vollenweider)—planned to write a sober nonfiction book. We became friends some thirty years ago at a seminar on the Age of Enlightenment. The differences between our academic disciplines baffled and perplexed us. Each of us became compelled with curiosity about the other's entirely different view of human existence in the universe.

Our discussions revolved around a basic question: Can today's science discover traces of God in the universe? Are there bridges from the old theological cosmology to modern physical theories about history and the structures of outer space? Or, to phrase the problem more succinctly, what might it mean in today's world of advanced science to talk about the universe as a divine creation?

After several false starts in the form of academic essays, we had to admit the impossibility of answering these questions with any single interdisciplinary theory. Physics and theology differ considerably in their understanding of reality and their methods. While struggling with how to accomplish our goal in a more flexible, user-friendly style, we came up with a special literary form to deal with this particular situation.

We decided that our book would not consist of a systematic or academic treatise but instead be the story of dialogues between a fictitious physicist, Sheldon Cutter, and a fictitious theologian, Thomas Haubensak, who each represents his academic discipline. They embark together on the long and dangerous space mission. After a series of unfortunate events, they are stuck together during long periods of waiting and must listen to each other, at length.

In our novel, a fictional team of a tech writer, an investigative reporter, and a sci-fi novelist are hired in the spring of 2051 to report generally on the *Hermes* mission. When contact with the astronauts is lost, they are assigned to produce a report based on all files that the onboard computer has transferred to the black box.

The two astronauts and editorial team—along with quotations from the media, expert responses, and narrative glue—are, of course, fictitious, though they allowed us, from time to time, to make a twinkling wink at life in today's scientific and theological communities. The discussions, however—along with the dialogue between Sheldon Cutter and Thomas Haubensak, personal journal entries, letters home, postcards for the Cyber School, and meditations—reflect an authentic engagement with our contemporary reality and science.

We would like to express our sincere thanks to Gwendolin Herder's great commitment on the side of Crossroad Publishing and to Alan Rinzler for his stimulating editing and other contributions; Christopher Bryan and John Gatta, for discussions; Linda and Wolfgang Thomsen, who made an immense effort in translating this book, and Emily Wichland for her patient copy editing.

We hope this adventure of a cosmic search for meaning will prompt a similar search in your life.

Arnold Benz and Samuel Vollenweider,
Zurich, Switzerland, Easter 2021

Crisis on the Elpís

On Monday, May 22, 2051, Harold Fender, the chief public relations officer of the International Aeronautical and Space Administration (IASA), sent out a signal to the three members of his team that summoned them on a video conference call. It was the day after Pentecost.

"We've lost contact with the astronauts!" he shouted.

"What?" Randall, the team leader, asked.

"You heard me," he repeated, his voice still rising and his face turning red. "First, the five astronauts on the ORPHEUS stopped transmitting, and now Sheldon and Thomas have blinked out on the ELPÍS!"

"Blinked out?"

As Harold spoke, he punched up a three-dimensional hologram of the Orbiting Radar Platform and Habitat for Excursions to Unknown Satellites, or ORPHEUS, and then the Emergency Lander and Prospector for Ice-Moon Studies, known as the ELPÍS (Greek for "hope").

"Mute! Nothing but gibberish radio signals!"

"OK, Harold," Randall said. "Calm down. Let's figure out what we should do."

Harold had recruited three journalists a few months ago to report for the wide international public on the *Hermes* mission to Saturn in the outer regions of our solar system. In the spring of 2051, the *Hermes Trismegistus* mother ship, or short *Hermes*, with seven astronauts reached Saturn. The interplanetary expedition, planned for a total of seven years, was dedicated primarily to explore Saturn's system with its spectacular rings and its moons Titan and Iapetus. The mission was named after Hermes, the Greek god of communication, who was combined in ancient Egypt with the god of wisdom and therefore called "Trismegistus," Greek for the "three times greatest": the greatest philosopher (scientist), greatest priest, and greatest king.

"The media is going berserk," Harold went on, his voice still shrill, his blood pressure rising. "Listen to this! The *Wall Street Journal*'s headline is flaming 'Tragedy or Ecstasy?'; the *Daily Express* in London says they've probably become captives of 'extraterrestrial beings'; *Fox News Interplanetary* is saying 'they're all frozen to death and dead as a doornail.' Referring to the astronaut Thomas Haubensak's spiritual competence, *El Mercurio* in Santiago, Chile, interpreted the last words received from the mission as 'glimpses into a transcendent world'; and the Finnish *Aamulehti* thought that they did find a giant hot spring and were enjoying a roll in the snow before plunging in."

"That's a good one," Randall laughed.

"It's not funny," Harold said.

"Sorry, sir. You're right," Randall said. "We need to get on top of this and send out something accurate."

Harold assigned the three journalists to produce a coherent and continuing narrative of the incidences within four days after retrieving the documents from the black box and computer memories of the space mission.

"I want a crystal clear account of what happened that's understandable to the general public. In particular, I want to understand the strange messages transmitted by the two cosmonauts during the last minutes of their descent to Titan."

The three journalists worked in different parts of the world and communicated in a chat room on the internet.

Hoihong Wong is an experienced tech writer and former engineer who lives in Singapore. She was responsible for producing accurate reports in language that the general public could understand. They'd decided that she'd use English, which Randall, the only native speaker on the team, would polish to the Queen's standard.

Astraia Callas is a crack investigative reporter based in London who was responsible for deep-throat data and analysis, without bias. Just the facts, but all of them. She was the detective on the team.

Randall Bradford is a science fiction novelist from Australia whose job it was to describe the characters and characteristics of every event with full dimension and literary skill.

Harold Fender is an old-fashioned American public relations flack, a former disciple of Louella Parsons, whose job it was as IASA's chief publications officer to stifle any damaging or idle gossip that would in any way tarnish the public's sense of confidence or infallibility of the astronauts and their mission. As a result, he'd developed the moniker Fend-Off Fender.

The journalists soon realized that it was essential to cover the preceding incidences and onboard discussions in April and May 2051 to understand the last stages of the mission. Thus their reporting starts in the spring, when the *Hermes* mission reached planet Saturn.

The journalists' team was well positioned for its task concerning the scientific activities aboard *Hermes*. They understood the flight and landing on Titan of ORPHEUS's five-person crew.

Unmanned probes, among them *Cassini-Huygens* launched in 1997, had already explored the world of Saturn quite thoroughly. This manned mission was now giving special attention to Titan, the only moon within our solar system surrounded by a dense atmosphere and reaching the dimensions of a small planet. Despite the severely cold temperatures, numerous organic molecules were detected and high-molecular chemical reactions seemed to occur, a precondition for the beginning of life in the early ages on Earth about three billion years ago.[1]

Although IASA discouraged any wild ideas about encountering archaic forms of life in this prebiotic laboratory, public interest caught fire at this point.

"Hermes Mission May Have Found Humanoid Aliens on Saturn" was the first of a series of headlines in the tabloid *International Scream* on May 15. Astraia Callas, the detective reporter in London, dug behind the scenes and was able to discover from anonymous sources in the *Scream*'s pressroom that the sales department had planned this campaign of bogus news to increase their circulation.

As a result, other objectives became totally secondary in importance to the general public, who only heard about the rest of the mission from the smaller, legitimate press. These other aims included an evaluation of a possibly exploitable, gigantic supply of hydrocarbons on Titan and ground-level liquid water on the ice moon Iapetus, both of which could prove useful in connection with the later installation of a permanent station.

An Unusual Crew of Astronauts

Despite the use of powerful ionic propulsion, the flight from Earth to Saturn took almost three years. In view of this enormous distance, the astronauts, unlike those involved in earlier flights to the moon and Mars, had to be able to operate independently of ground support systems. Therefore, IASA had subjected the astronauts to exceptionally rigorous and prolonged aptitude tests by an international recruitment committee.

Seven young, technically well-trained scientists were finally chosen: **Sheldon Cutter**, an American physicist with a doctorate in planetary science and several years of experience on a small lunar base. He is the leader of the expedition team. Pilots are **Takeo** from Japan and **Sergei** from Russia. **Shadia** is a Palestinian computer scientist, **Pablo** an African American doctor. Members of the team are also **Nicole**, the French biochemist, and, surprisingly, Thomas Haubensak, a German theologian.

Thomas Haubensak was recruited as the result of IASA's endorsement of a religio-psychological research project proposed by the society for Participatory Existential Perception (PEP) in San Francisco. They determined that this astronaut was assigned to test the numerous reports about alleged religious experiences in outer space that had been circulated since the beginning of astronautics, particularly the sixth man to walk on the moon, *Apollo* astronaut Edgar Mitchell. Mitchell said that on the trip back to Earth, he felt a profound sense of universal connectedness with the stars, the receding moon, our blue planet, and the vastness of the cosmos.

The presence of divinity became almost palpable, and I knew that life in the universe was not just an accident based on random processes. The knowledge came to me directly—noetically.[2]

Mitchell founded the Institute of Noetic Sciences (IONS) in 1973, an antecedent of PEP, to create a multidisciplinary field of study that brings objective scientific tools and techniques together with subjective inner knowing and spirituality to study the nature of reality.

The IASA's scientific committee expressed vehement objection to the idea of including Haubensak, protesting that he was a pseudoscientist and negotiated his role to compel a reformulation of the candidate's profile with the assignment of additional duties. The theologian Haubensak, trained and educated also in the humanities, was now to bear an additional appointment as group therapist and specialist for arbitration, since major interpersonal and psychological problems were expected on this long voyage. A conflict among crew members is one of the worst things to happen on a space mission. But to the annoyance of the finding committee, the American press—particularly the *New York Times* and *Washington Post*—took excessive interest in Haubensak and the many hats he wore as the mission's theologian and psychotherapist.

Finally, the astronauts reached Saturn's system. The mother ship, the *Hermes Trismegistus*, swung via a complicated orbital maneuver into a waiting position at the Lagrange point between Saturn and Titan, where the gravities of both celestial bodies oppose each other and maintain a balance with the centrifugal force.

So, *Hermes Trismegistus* locked on to the moon and circled with it around the gigantic gas planet in sixteen days. On March 31, five of the crew members—Takeo, Sergei, Shadia, Nicole, and Pablo—started toward Titan with their descent module ORPHEUS to erect a ground station, while Thomas and Sheldon remained on the mother ship.

The landing of ORPHEUS was delayed and the contact between *Hermes* and mission control on distant Earth suffered badly from communication disturbances. This sudden radio silence left the two remaining crew members on *Hermes* without much direction or support from the IASA command center on Earth.

Thomas Haubensak and Sheldon Cutter had good radio connection to their fellow astronauts who had descended toward Titan to explore its surface. Ultimately the landing crew found themselves in a dangerous situation because of a defect in the primary energy supply. Thomas and Sheldon decided to descend with the second landing craft, ELPÍS, to Titan on May 20, without awaiting approval from IASA.

Then IASA lost contact with Thomas and Sheldon on ELPÍS, which is when Harold Fender called the journalists' team.

Nevertheless, the automatic sound recorders kept running in the *Hermes* black box, so everything they said out loud to each other on *Hermes* was recorded for facilitating the investigation of onboard incidents and accidents. In addition, any sound in the ELPÍS landing craft was transmitted to *Hermes* and saved.

The Records of the Conversation

On June 16, 2051, ground control made the decision to copy and transfer all information from the black box and all computer memories of the *Hermes* in order to learn more about the unexpected fate of the Saturn expedition. The three journalists of Harold's press team had the first and fullest access to these recordings and nontechnical documents, long before any of them were leaked to the media or given to so-called experts to dissect and analyze.

Our three-person editorial team listened to all of the recordings in the black box of *Hermes*. They commented to each other by audio chat on the internet and asked Harold to monitor their work as they listened to the sequence of the files. They worked nearly full time for five days, from June 16 to the deadline on June 20, so the different times zones did not matter.

They found many interesting dialogues at dinnertime in the "Cupola." Both astronauts liked to use their free time cooking food items that might seem unusual, but they mostly combined foodstuffs grown onboard with highly concentrated provisions. The meal was frequently accompanied by animated, in-depth conversations.

The Cupola, a cabin with many windows, rather extravagantly arranged in this otherwise very functional spacecraft, constituted the galley, dining room, and lounge. The dark paneled floor, the light walls, the energy-hungry but cozy halogen lamps, and the dome-shaped panorama window in the ceiling gave the crew the feeling for "above" and "below." The crescents of Saturn and Titan were visible through the window. When the lights were turned off, thousands of stars shone still and magnificently.

Meanwhile, the two astronauts floated in total weightlessness, in any position they chose, like embryos in the womb. Usually they held on to a foot grip during meals, so they wouldn't float away from what they were eating.

"No-gravity eating can make a real mess," Hoihong Wong, the engineer journalist commented. "I've tried it!"

The air was pleasantly warm, so the spacemen could wear loose-fitting clothing. The relaxed eating position and the open view into outer space certainly helped to inspire their conversations.

To while away the time, these two—the theologian and the physicist—had occupied themselves with, among other things, challenging conversations that led to a dialogue between the two astronauts. Each came to this mission with such different educations and consequently distinct and separate core beliefs. Topics for their debate included whether it was the God of the Judeo-Christian Bible or the big bang that created the universe out of "nothing," the creative presence of God in the cosmos, the question of randomness or design in the evolution of life, the uncertainty and contradictory behavior of matter and energy at the subatomic level in the new quantum physics,

Figure 1. In December 2010 a Large White Spot of ammoniac ice crystals formed on Saturn. The *Cassini's* infrared telescope's photo shows it as a gigantic hot spot at its maximum extent of a diameter approximately three times that of the Earth. (Photo: NASA/JPL-Caltech/Space Science Institute)

the dimensions of time, the meaning of human awareness in outer space, and possible scenarios of the distant future of the universe.

What follows is the word-for-word transcript of the debate between Sheldon Cutter and Thomas Haubensak, including Sheldon's "Postcards from Saturn for the Cyber School Program," Thomas's "Night Diaries," and some other essays and quotations that the journalists thought were relevant and were inserted at the appropriate places.

To finish the job that Harold Fender had assigned the investigating team, the recordings of what happened during the last three days of the mission to Saturn are added at the conclusion of this document.

Creation Today:
An Easter Debate about a Hurricane on Saturn (April 2, 2051)

SHELDON: That white hurricane on Saturn's surface fascinates me. What a spectacle! Every ten hours the rotation of the planet brings it into sight, and I greet it like a touch of spring in the winter's cold.

THOMAS: Yes, we've got an excellent box seat up here. That violent storm must unleash tremendous power. To me, too, it seems like a seed of life in a world of death.

SHELDON: It's amazing how fast the storm built up before our eyes!

THOMAS: You're the physicist, Sheldon. What causes this hurricane in Saturn's atmosphere to arise so unexpectedly? What's the scientific explanation for how a new structure appears out of nothing?

SHELDON: OK, my theological friend, I'll explain it to you. On this planet there are no oceans, mountains, or volcanoes. In fact, it has no real surface. Its hard core in the center is minuscule. Saturn consists of practically nothing but gas—a gas that's actually more uniform than Earth's air. Nevertheless, the Saturnian atmosphere is not a closed system in equilibrium. The inner part of the planet is much warmer than the outer layers. This amounts to an enormous energy potential, into which small random currents in the atmosphere can tap and intensify themselves. Are you following me?

THOMAS: Yes. Yes. So far. Go ahead.

SHELDON: OK. So gas rich in ammonia is flung out from the depth to great heights and freezes there to become ice crystals, forming a white spot much like an earthly ice cloud. This way, a tiny initial fluctuation in the swaying of thermal agitation becomes stronger and stronger, and a vortex forms on its own. Finally a hurricane emerges—a typical example of self-organization in nature.[1]

THOMAS: Wow. I'm astonished that a seemingly dead cloud layer has the ability to bring about a large-scale tempest quite spontaneously. Ordered structures evolve out of a seething sea. And not even a physicist like you can predict this!

SHELDON: Hold on, Thomas. Yes, science recognizes that tomorrow can be totally different from either yesterday or today, but the laws of cause and effect always apply, even if we haven't discovered yet how the process works. For example, we didn't know back in the twentieth century that structures can form out of thermal disorder when they become self-augmenting through feedback, like this white spot on Saturn, which is a good example of that.

THOMAS: A new order arises out of chaos! So creation is not limited to the beginning but continues to unfold today!

SHELDON: No, no. Nothing is "new" in the world of physics. These violent storms happen to develop occasionally on Saturn. But they don't emerge out of nothing. Not a single atom is new, no additional energy appears. It's just a reorganization stimulated according to the predictable science of existing elements.

THOMAS: I don't see it that way in my world. I don't agree that everything was already here before.

SHELDON: I suspect that you're chasing after the new because you're looking for the hand of a creator God.

THOMAS: You're right. When I asked about the origin of the new, I was wondering if there's any point of contact here between physics and religion.

SHELDON: Nope. The new that science observes needs no special conditions, no external powers, and certainly no supernatural factors. The course of causality isn't interrupted. And when chance plays a part, its operation is random.

THOMAS: I'm not looking for inexplicable miracles. New and wonderful things can happen in the altogether "natural" course of events. A child is born. A person experiences unexpected rescue from a deadly peril or new life after a personal catastrophe. There are days when one sees the world as on the first morning of creation.

SHELDON: We're not talking about the same thing. The newness you have in mind is totally subjective.

THOMAS: That's precisely the point! You admit that in a strictly physical sense, there is nothing new, nothing that never existed before—or at least, not in our experience. But I'm saying that the very concept of newness suggests a unification between subjective and objective perceptions of reality. That's what I find so intriguing and attractive about it. But let's return to this topic later. Now I'm eager to see what new things await us in EDEN, the Experimental Division for Ecology and Nutrition, the bio-zone of the spaceship.

SHELDON: Yes, I'm eager to see if we can discover something new among the plants on this artificial life oasis far from Earth.

THOMAS: Yes, and this would be a good time for a festive meal, since today is Easter Sunday on Earth. A fascinating association between Easter and our talks about the hurricane on Saturn keeps running through my mind.

SHELDON: What on earth—or out of it—makes you think of that?

THOMAS: Easter's message is about life that emerges out of death. From time immemorial, the transition from Good Friday to Easter morning, from death to resurrection, and from darkness to light was an essential part of the Easter symbolism.

SHELDON: Sounds pretty vague. Are there any objective facts?

THOMAS: The gospels tell that the followers of Jesus Christ fell into a deep crisis when their master and teacher died on the cross. But according to their statements in the gospels, Jesus rose from the dead. Life, therefore, has irrevocably overcome death.

SHELDON: The evidence seems to me somewhat meager to say the least. Our ARGOS telescope can look into the past of our universe but not, unfortunately, that of Earth.

THOMAS: You're right. The Easter story has no objective data. But it testifies to intense experiences that shaped the future life of these people and ultimately millions of others all over the earth. Such a practical consequence is a vital form of reality in every sense. The scattered flock became a living community, and hope arose from despair.

SHELDON: What you're describing, though, are subjective experiences of an ancient sect. It could all have been an illusion.

THOMAS: Religious experiences simply elude your criteria for objectivity. But this is also true for many other human experiences that depend on a people's perceptions of reality. Music, literature, dance, opera—different critics and patrons of the arts have disagreed mightily in their judgments of art over the centuries. Mozart had "too many notes," Picasso was a "hack . . . a plagiarist," Tolstoy was "too long." So art, like religious belief, reveals itself as a reality when people share certain experiences with one another. Opportunity for such sharing remains available to us today.

SHELDON: Wait a moment. I just heard a sharp ding from the onboard food warmer. The main course of our cuisine must be finished warming up.

THOMAS: Right. I'm still getting used to eating without cutlery to keep crumbs from floating away in this weightless situation. It's awkward to press the food and drinks filled in small plastic bags directly into my mouth. But this broccoli isn't bad, don't you think, Sheldon?

SHELDON: That's one thing we can agree on.

THOMAS: And the chicken?

SHELDON: Not so sure.

THOMAS: Very funny. Did you learn a lot about chickens back where you come from?

SHELDON: Not a bit. I grew up in West Fargo, North Dakota. But we all learned in the fourth grade about the most famous young guy in the city—astronaut Anthony England.

THOMAS: I've heard of him.

SHELDON: He was the youngest man ever to become an astronaut. He flew for the Space Shuttle Program, logging 188 hours out in space.

THOMAS: That was quite a big deal back then.

SHELDON: Yes, it inspired me to become a pilot and then to become an astronaut myself. I had to struggle to get higher education, since I was the oldest of four kids in a poor working-class family. But I had the best grade of anyone in high school physics, and I won a scholarship to North Dakota State in Fargo. Then I earned a doctorate in planetary science at the University of Alaska Anchorage.

THOMAS: That's a circuitous route, with some bumps in the road.

SHELDON: Ouch. You're right, but I began flying small planes in Anchorage, then I joined as a pilot in the International Pacific Rescue Service. I made a hasty early marriage that broke up, so I switched to IASA and had myself stationed on a small moon base. I loved life on the moon, so at thirty-six years old, I tried out and won a spot on the *Hermes* mission. So now you know all about my struggles to get to this point.

THOMAS: So here we are.

SHELDON: Like peas in the pod—trapped together.

THOMAS: Ha! Good metaphor.

SHELDON: But I don't know much about you yet, Thomas, except that with a name like Haubensak, I'm guessing you're German, so . . . *bitte mit mir reden.*

THOMAS: You speak German?

SHELDON: No, sir. I memorized just those words when I saw your name. Your English is a million times better than my German, so please tell me all about yourself, Thomas.

THOMAS: Very well. I was born in Berlin forty-three years ago, so I'm five years older than you, young man. My father was a doctor, but I didn't follow his path.

SHELDON: Aha, a rebellious youth!

THOMAS: You could say that, I suppose, but more interested in our minds and spirits than revolution of any kind. I studied theology, classics, and religion at Tubingen and Zurich, then I trained as a psychotherapist in Pasadena, California. Nowhere near North Dakota, but your country nonetheless.

SHELDON: Why did you come to America when Germany and Austria are the cradles of psychoanalysis?

THOMAS: I wanted to get away from Germany and see what it was like in a new, diverse society with so many different kinds of people.

SHELDON: Interesting. And do you have a family.

THOMAS: Yes, I've been married to my beloved wife, Lilian, for almost twenty years. We have one daughter, Margo, who is now sixteen, and a son, Erik, who's twelve.

SHELDON: Wow, congratulations. It must be tough to be away from them for so long and so far.

THOMAS: Yes, quite . . .

SHELDON: Look! The sun is rising again like a sparkling diamond in the fine blue layer of haze above the crescent of Titan. As you said, we come out of darkness into light. But I still can't understand what kind of bridge you want to make between Easter and the tempest in Saturn's atmosphere.

An Eighth Day of Creation

THOMAS: What if you were looking at this storm as if it were an experiment, formed under the extraordinary conditions found in the realm of Saturn! Doesn't this pattern emerge again and again in those natural processes where, in the midst of growing turbulence, a new and relatively stable state suddenly establishes itself?

SHELDON: Yes, but . . .

THOMAS: Well, for me, Good Friday and Easter are like the universal symbolism of death and regeneration. And the message of resurrection is about something fundamentally new that was never there before. For Christians, the resurrection of Jesus becomes the focal point of a new event of creation occurring in the midst of history. Easter marks the dawn of the eighth day of creation![1]

SHELDON: The eighth day? I get it that Easter seems to be for you something like a classic instance of God's creative activity long after his mythical creation. But as an empirical scientist, I'm not satisfied with only one data point. Can you tell me some other examples?

THOMAS: I can. Easter isn't an isolated case. The history of Jesus condenses the whole historical drama of the Israelites. The Old Testament testifies how God, especially in critical situations, repeatedly allowed himself to be newly perceived. Thus he led the Israelites out of Egypt into an uncertain future through a barren desert. Later, God's people, amid the miserable hardship of Babylonian exile, are surprised with a decree of the Persian king that enables their return home.[2] The prophets confront Israel with its imminent downfall while trying to open its eyes to the life and fresh possibilities that God offers in a totally unexpected and new way. And did you know that Saturn, our host, was regarded in antiquity as the star of the Jews?[3]

SHELDON: I didn't. But all that has nothing to do with the emergence of new things in nature.

THOMAS: You're right. The biblical tradition perceives newness primarily in the field of human history. But the biblical tradition also recognizes in that process the work of the same God who called the world into being and preserves it. The rescue of the Israelites from Babylon was celebrated as a new creation.

SHELDON: Yet even if certain historical events may be compared to steps in the development of the universe, the question remains: What's the point of stretching religious conceptions to describe patterns in nature? After all, the success of modern science lies chiefly in its capacity to interpret nature without drawing explanations from mythology or mythological patterns. Most of the population in Jerusalem didn't notice anything worth mentioning on that day, aside from the disciples of Jesus.

THOMAS: Right. Looking at the matter historically, from the standpoint of documentation apart from the New Testament, we know almost nothing about the Easter experiences of Jesus' followers. But this much can be inferred: during the passion and crucifixion of Jesus, they fell into a deep crisis. Everything seemed to break down for them. They returned distraught to their homes in Galilee. But then something quite upsetting and momentous took place. Many of them were surprised by a unique vision in which Jesus is revealed in a body of light, shining in heavenly magnificence. We don't know what really happened. The fact is that the disciples felt a presence of Jesus as if he were living. This totally changed their perspectives and their lives. They were jubilant and motivated to spread that news all over the world. The room of death, formerly bolted so hopelessly, had now been opened.

SHELDON: So you don't proceed on the assumption that Jesus departed in bodily fashion from his grave after his crucifixion, but instead you speak rather cautiously about a vision. Let me tell you a story, Thomas. Many years ago, I had a fierce discussion with a devout believer about whether a truly dead person could return physically to our world. Above all, my discussion partner back then couldn't explain to me where this person traveled to upon his presumed ascension to heaven. As a future astronaut, I wanted to know something more precise.

THOMAS: At this point you're kicking at an open door, Sheldon. A video camera couldn't have filmed the resurrected Jesus. We're not talking about a stream of photons from a body of light. The extraordinary phenomena were a special kind of vision.

SHELDON: Visions?! Here's where I have a problem. Why are hallucinations given such authority to define the beliefs of an entire world religion? The notion of "building structures out of chaos" could be just as well attributed to psychotic conditions. I have heard that hallucinations can be a defensive reaction. Did the disciples of Jesus handle their crisis with the help of projections? Couldn't they be post-traumatic delusions?

THOMAS: I prefer to talk about visions instead of hallucinations, even if, from an external perspective, it's impossible to distinguish them. A true vision confers a kind of perception in which an otherwise hidden dimension of reality is unveiled. A vision involves the opening of an inner eye, therefore leading to a higher state of consciousness.

SHELDON: Inner eye? You're getting further and further out, my friend. This kind of differentiation relies on highly questionable terms and assessments. I can't do anything with "inner eye," or "higher," or with the "concealed dimension." For me, visions and hallucinations are the same.

THOMAS: The subsequent life of the early Christian may give us some helpful criteria to understand their differences. Visions can deeply influence people's experience and behavior, while hallucinations rarely produce major, long-term consequences. "You will know them by their fruits!"[4] Moreover, higher states of awareness are not organized around a defensive psychic pattern, as is usual with the mentally ill having a hallucination. Visions allow access to experiences unavailable within ordinary states of consciousness. They occur in a psychic state that is not a rigid structure but remains fluid, dynamic, and open-ended.

SHELDON: That's right! The extreme yet monotonous external conditions here in outer space sometimes put me in strange moods, too. The astro-psychiatric trainers often warned us about the dangers of perceptual disorders and loss of reality.

THOMAS: Being out here in space certainly makes us more sensitive to the possibility of other states of awareness and liminal experiences. But we can't relegate the question of whether such experiences are pathological or real perceptions solely to psychiatry. Religion can't be reduced to psychopathology.

SHELDON: Why not? You yourself mentioned visions and bizarre states of awareness.

THOMAS: True, but religion doesn't depend primarily on these visions and higher states of awareness on a daily basis. It gives us orientation, meaning, and a moral compass in our daily life. It gives us the occasion for loving connection on collective occasions: holidays, important life-cycle rituals such as birth, coming of age, anniversaries ... death. Faith gives meaning and inspiration to our lives. And all of this integrative potential is informed, though, by certain experiences of heightened awareness related to prophecy and mysticism.

SHELDON: All of us—believers and nonbelievers alike—have had experiences where we become aware of a harmony echoing in the entire cosmos. They bestow a deeper meaning to one's own limited life. For me, this feeling for the beauty and dynamism of the universe is an elemental driving force for my scientific work and my long excursions into space. It appears to me that you understand the visions of Jesus' disciples in a similar way.

"What in the world is Thomas talking about?" Hoihong asked. "I would very much like to understand it as a tech writer and former engineer."

Randall jumped in: "Same for me. I have asked an expert on the New Testament to comment on the dialogue regarding the interpretation of Christ's resurrection. Here's what he answered."

Assessment on the Question of Jesus' Resurrection—Reality or Vision?

The theologian and the physicist are debating whether the resurrection of Christ is an authentic miracle vision or a hallucination. They are not, however, speaking of the same thing. The so-called vision hypothesis came into popular currency only in early modernity. Critics with a more classical attitude put forward at least three main arguments against this hypothesis:[5]

First, the early Christian texts explicitly speak of Christ "being raised from the dead" or of his "rising from the dead" but not of "visions." And this despite the fact that in antiquity, many claimed to have had visionary experiences and such experiences were frequently reported.

Second, the alternative between "material resurrection body" and "spiritual reality" is misleading. In Judaism and Christianity at that time, there were many different ideas about the nature of the resurrection body; the scale ranges from a restored material body to a kind of spiritual aura or aura of light. Thus in the view of the gospels, Jesus' resurrection body is on the one hand similar to his earthly body: the resurrected person is no longer in the tomb; he eats with his disciples and lets himself be touched by them. On the other hand, his resurrection body differs from the earthly way of being: the resurrected Jesus walks through closed doors; he sometimes appears in a different form so that his followers do not recognize him. In this respect, his body is (in Paul's words) a "spiritual" body (1 Corinthians 15:44); he has experienced a transformation, a "transfiguration."

Third, talking about someone having risen from the dead was to make a claim that most people in the world outside of Judaism would have found either offensive or ridiculous. Had the early Christians merely claimed that they had seen visions, their contemporaries in general would have found that much easier to understand and accept.

Thomas Haubensak would certainly present arguments against these three points of criticism. He would emphasize that the ancient Christians were firmly convinced that Jesus physically rose from the dead. But this conviction already represents a certain interpretation of the experiences made by the Easter witnesses. And that the reported experiences were mostly of a visual nature. "The Lord was indeed raised from the dead and appeared to Simon" (Luke 24:34); he "appeared" to several witnesses (1 Corinthians 15:3–5). The modern reference to the video camera, which in this case would not have recorded anything, is trivial but inevitable. Thus the classical interpretation is not understandable in today's worldview shaped by science.

A third way to understand resurrection avoids the notions of material reality and purely internal visions. What the Bible reports is a much broader class of experiences known as religious experiences. They are not scientific measurements or observations, neither objective nor reproducible. They are also not purely subjective, because the disciples participate through feelings, recollections, and even bodily. Art, human relationships, and many more experiences in daily life are also perceived in such a participatory way. The truth of such perceptions cannot be objectively verified, but it must be confirmed in its effect. In the case of the resurrection, such confirmation could be found in the radical change that the Easter experiences caused in the lives of the early Christians or that the message of the resurrection spread astonishingly fast and effectively all over the world.

Against the background of these arguments, it becomes clear that both the physicist and the theologian onboard the spaceship use an ambiguous term, *visions*, to understand Easter, which led to their misunderstanding each other. The scientist reckons that the disciples could have been suffering from hallucinations, while the theologian interprets their reports as participatory perceptions having a crucial relation to reality.[6] It may be better to avoid the term *vision* altogether.

—Professor Kevin Nachtigall, Institute for Interdisciplinary Studies in Biblical Research, Montreal

"I have the impression that Thomas has a rich inner life and participates emotionally in all that's around him. Let's have a look at his diary of that night," Randall suggested.

Figure 2. The *Cassini* space probe took this photo of Saturn from the vicinity of the moon Titan, which is visible on the left. Titan is covered by an opaque orange-colored aerosol. Saturn's rings can be seen edge-on as a thin horizontal line in the middle of the picture. The rings cast a shadow on the planet. The A, B, and C rings (from top to bottom) are separated by bright divisions. (Photo: NASA/JPL-Caltech/Space Science Institute)

In Saturn's Sphere
From Thomas's Diary (April 2)

I have just worked my way, hand over hand, from the observatory to the main module of the spacecraft. Only the humming of the ventilation system and the occasional sound of radio communication with the Titan landing craft ORPHEUS *break through the deep silence.*

Titan shines through the porthole, a seemingly motionless orange ball more than ten times as large as our earthly moon when seen from Earth. It's shrouded in a thick atmosphere, offering the human eye no visible structures. What kind of secrets might the expedition find there?

I'm hardly able to spot the distant Earth—our home. The sun seems to be only a distant bright light bulb in the deep dark night.

The globe of Saturn decorated with its rings shines through the opposite window. It appears smaller than the nearby Titan, but still huge compared to the moon seen from Earth. The ring system composed of millions of ice and rock fragments offered an impressive sight at our approach. Now we see it edge-on and just like a thin line. Since we circle in the same plane as the rings and moons, we've become truly part of Saturn's system.

We have distanced ourselves far from Earth, that island almost overflowing with life in the midst of a yawning emptiness that is apparently hostile to life. I'm astonished at how life nested and lodged itself in those early days on Earth, a remote niche in deadly cold space; and how life has ceaselessly persisted in fighting death for its fragile existence, constantly threatened with extinction. Lifeforms flash up on Earth like sparks amid the cosmic abyss of space and time.

I can hardly believe that we linger now in Saturn's sphere, in that region that, for antiquity and the Middle Ages, formed the border zone between the planetary world and the fixed stars. In those days, anything farther away was believed to be the "Empyrean Heaven," occupied by fire or ether in Aristotle's natural philosophy.

Saturn—the guardian of the threshold and principle of matter, master of death and angel of destiny, gatekeeper over time and laws of karma! Beyond this threshold, the sphere of the highest God beckoned for humans of ancient times. In our days, we reel in view of the sheer vastness and immeasurability of space outside our solar system, the star clusters, and the galaxies. The eternity of heaven has given way to the immensity of space and time.

I am struck by recollection of the checkered history of Saturn, the Greek god Cronus. He had snatched away domination of the world from his father, the heavenly god Uranus, but then together with his siblings, the Titans, was deprived of his power by his own son Zeus and incarcerated in the dark Tartarus. According to another ancient Greek tradition, he now reigns over the Islands of the Blessed, a place like Paradise:

Beside deep-eddying Ocean
—happy heroes, for whom the grain-giving
field bears honey-sweet fruit
flourishing three times a year.[7]

Ancient tradition attributed a double nature to Saturn, the Greek Cronus. On one side he reigned as dark ruler of the sky, old and melancholic, representing fate and the principle of retaliation. An old Gnostic teaching regarded him as nothing less than the star of entropy:[8]

No one fixed in birth is able to escape Cronus's power in the whole world of generation, Saturn is responsible for subjecting things to death, and no birth occurs in which Cronus does not interfere.

On the other side, Saturn was regarded as the patron of the philosophers and prophets.[9] *For ancient observers of the heavens, he was the farthest away, but also the slowest planet, dignified and solemnly revolving in his orbit around Earth. Since his sphere was the highest, the philosophers sought in him the spirit of magnificent vision and perfect wisdom. He was placed above Jupiter, the star of active royal sovereignty. The secluded thinkers and hermits were the children of Saturn. Not for nothing was he the patron of the Platonic Academy of Florence.*

Out here I can readily understand and marvel at the double nature of Saturn. A terrible storm rages on this ice-cold giant planet. It announces from afar a surprising creativity, some tentative movements of life to come. Death and life meet in its sphere.

"Look what I found," Astraia exclaimed. "Later in the evening, Thomas wrote a letter to his wife, Lilian."

"Hmm ... the text is rather intimate, and it is against our rule to publish personal writings," Hoihong objected.

"What do you think, boss," Randall asked Harold.

"Let's ask Thomas's wife for approval," Harold replied. "While we wait for her reply, we can go ahead and quote some passages that illustrate the mood that's behind these night diary texts."

"OK," Astraia said. "Here's what he wrote."

My love,
It is three years since we said goodbye to each other shortly before the start of our mission ... Sometimes the loneliness out here in the outer reaches of our solar system grabs me with ice-cold fingers. Then all our psycho training, for which I am especially responsible onboard, is of little help. We have been thrown into a vast ice-cold desert. My thoughts keep returning to our oasis, to our wonderful Earth.

Actually, I am convinced that life and intelligence are not limited to Earth but scattered throughout our vast universe, and that it is only a matter of time before we get in touch with aliens. But in these melancholic moods I no longer trust my own thoughts, and I ask myself, What would it be like if we were actually alone in the universe? If the pessimists were right?

Then Earth would really be a tiny and absolutely unique biotope in the immeasurable cold abysses of the universe. And you, my dear wife, are the heart of this mini biotope. I miss you and Margo and Erik so much. As grandiose as the world of Saturn presents itself to us: We are not made for this cold glory. I'm homesick and am longing for you, for our partnership.

I will love you forever, your Tomi

SHELDON: This morning, I'm going to put the hundred-foot Arrayed Reflector Gadget for Optical Studies—or ARGOS telescope, as we call it—into

An Eighth Day of Creation

operation for the first time. The telescope was detached from the HTM for the observation to avoid vibrations caused by our movements.

THOMAS: How does ARGOS work?

SHELDON: It collects the light in twenty-seven small concave mirrors that are attached in a Y-form and together produce the same magnification as a telescope with a diameter of one hundred feet.

THOMAS: And what are you pointing it at?

SHELDON: At that Great White Spot. According to older reports in the astronomical literature, Great White Spots in Saturn's atmosphere occur about every thirty years.[10]

THOMAS: As horribly as the giant storm may rage on Saturn, I'm fascinated by its beauty from our safe viewpoint. Since the Titan expedition has left, it's become rather monotonous here. It's an unexpected gift that we can observe these Great White Spots at such close range.

SHELDON: The storm is about three times the size of Earth. Its energy amounts to the annual energy production of one million large electrical power plants on Earth. The wind speed at some points exceeds four hundred miles per hour. Clouds are formed continuously from different molecules, ice crystals or drops, and then disintegrate again.

THOMAS: I'm astonished that something so big can form repeatedly according to the same simple laws of the smallest particles.

SHELDON: Same with me! Perhaps our feeling has to do with the complex order of this storm. Elementary forces act between numerous particles; their interactions produce a self-organizing system. Instead of a chaotic jumble, it becomes a whole.

THOMAS: I think it's remarkable that you're talking about "whole" and "complexity."[11] Usually I have the impression that the sciences look for ways to break down the complex into the simplest and smallest possible pieces.

Don't physicists want to derive everything from basic laws and elementary proportions?

SHELDON: Yes, but the scientists who discovered physical laws over the centuries have always been astonished to see how elegant nature is. The fundamental equations of physics are surprisingly simple and small in number. We still suspect that there may be a single world formula. One could conclude from this that the universe is also simple, comparable to a mechanical clock with its gearwheel mechanism. But not at all! The implications in nature can be far more complicated than its laws. Those fundamental cosmic laws permit an immense number of surprising outcomes, giving the universe a gargantuan capacity for development. Accordingly, structures are constantly re-forming into new objects of an ever-higher order.

THOMAS: Science and religion also seem to meet in strong astonishment. I'm not thinking about miracles or extraordinary events here but rather the most elementary things that have filled religious people with amazement: the gift of life, the abundance of being, and the overwhelming dimensions of the world. One could describe religion as a culture of handling the astonishing. But here's an important point, my friend. In contrast to science, religion doesn't work toward explaining amazing things and thereby nullifying the astonishment.

SHELDON: Science doesn't nullify astonishment. Each advance in knowledge provokes further questions, generating new surprise and inspiration. Above all, amazement captures researchers about phenomena they *can* explain. The more I understand, the more I'm astonished! I'm impressed, for example, at how exactly we can calculate the short-term movements of celestial bodies. Without this knowledge, we wouldn't have found Saturn's system in the vastness of interplanetary space. It also amazes me how many phenomena of this bizarre region can be correctly forecast. And I don't know what I should admire more: the mathematical conformity we have found with basic laws in nature or the ability of our minds to recognize this.

THOMAS: Just a moment.

An Eighth Day of Creation

SHELDON: What is it?

THOMAS: We just got a call from ground control for an orbit correction. The *Hermes Trismegistus* has drifted too far from the Lagrange point.

SHELDON: OK. Quick! We have to ignite the propulsion units for a calculated thrust . . .

THOMAS: All right . . .

SHELDON: Done.

THOMAS: Good work. But just before that interruption you said that you and other scientists were continually astonished when they study and learn to explain phenomena. To me, however, your astonishment remains something subjective and apart from knowledge. You and your colleagues seem inclined to mask the affect that often accompanies discoveries. In religion, however, wonder is affirmed as critically meaningful. This astonishment discloses a hidden dimension of reality, a kind of resonance with the world's secret and its divine foundation. The knowing is inseparable from the way that leads to it. It is affective wonder that vitalizes religious statements and provokes ongoing discussion about them. Talking about God as the creator of the universe stirs up the feeling of wonder. A religious statement that doesn't elicit this attitude remains incomprehensible. By contrast, a mathematical equation need not inspire any affective response. One can understand it without inner emotion and sympathy.

SHELDON: So you're saying that only religion can do justice to wonder. I totally and emphatically disagree. Wonder and curiosity also guide and stimulate science. In fact, science actually does far better at sustaining a key precondition of amazement: the openness to questioning. With its ready answers, religion prevents this openness.

THOMAS: No, not at all. Religion is all about unlocking the door to a wide room and nurturing our awareness of a divine presence in the world.

SHELDON: Do you want to claim this even for the Great White Spot?

Figure 3. Old Saturn with crutch and his murderous sickle. (Codex Schürstab, Nuremberg, about 1472, p. 25v, Zurich Central Library)

THOMAS: Yes! I'll never forget the sight of the seething white hurricane-storm on Saturn. Something of the secret of creation flashed up. The wonder calls for me to be thankful. For religion, this richness expresses the downright extravagance of divine creativity.

SHELDON: Hmmm . . . interesting . . . But in my view, there is nothing holy about highly organized complexity. Our spaceship's control systems are very complex. The genetic material of a deadly virus is no less complex, considering the number of its autonomous components. Are you astonished by those, too? Certain phenomena arouse amazement in us, but others indifference, fear, or disgust. I think science does well not to incorporate any subjective responses into its portfolio of knowledge.

From the Old Cosmos to the Modern Universe

THOMAS: There's another aspect to the subjective response, Sheldon, that's impossible for even a scientist like you to ignore. The sight of the hurricane's order is impressive from our safe spacecraft. But how dreadful if such a storm, guided by the same laws, raged on Earth, a planet filled with life! I'm sure you'll agree that wonder would change into horror, praise into lament, thanks into curses.

SHELDON: Can't argue with that.

THOMAS: And here's something else I'd like you to consider. In some ancient religions, celestial bodies with luminous nature were attributed divine status. Now the status of these grand objects, at least with regard to their order of complexity, are degraded below that of a hairy little amoeba! Judeo-Christian tradition, however, assigned special dignity to the *human* form of life. When God created Adam, he demanded that the angels, clothed in a body of fire, prostrate themselves before this creature made out of soil. Satan's refusal to do so is said to have led to the fall of angels already in the morning of creation.[1] Even later on, the angels born of light were jealous of humans, privileged by God. Of course, divine choice was thought to be a decisive determinant of Adam's standing. Today we might esteem bacteria and protozoa, because of their more complex structure, over the unimaginably larger objects of stars and planets.

SHELDON: Nice story. May I work on it a bit more? Our organism, a body formed out of dirt, which the angels despised, is a good example of cooperation between local, accumulated order and globally increasing disorder. During our discussion, cognitive processes have, on the one hand, created mental order, while, on the other hand, the body has changed substances like the ambrosia fruits into amorphous matter and giving off heat to the world around us. It means that every increase in order has to pay its price. The price consists in the global increase of disorder and decay.

THOMAS: I wonder whether Jesus' death on the cross and his resurrection might suggest an analogy to this elemental cosmic law. It's actually death that makes the resurrection possible. The complete decay symbolized with the cross enables a new order to arise, a higher complexity, symbolized by the resurrection.

SHELDON: How can you blend together an increase of entropy with the cross? People in the ancient world hadn't the faintest idea about modern physics. And what does an execution have to do with entropy? You're confusing unrelated things here.

THOMAS: Of course, the theologians of the early church weren't acquainted with modern thermodynamics. For them, the cross was an instrument of torture, a terrible device for bringing pain and death. Christ endured pain on behalf of all humankind. Sometimes, in fact, Christian tradition has seen Jesus' suffering as typifying the suffering of all creatures, including the animals. In the cosmological symbolism of early Christianity, the cross even anticipated the downfall of the old world. A worldwide darkness and the earthquake at Jesus' death were signs for the impending apocalypse.[2]

SHELDON: But you're nowhere near to connecting crucifixion with the rise of entropy, which is a physical manifestation.

THOMAS: Perhaps we've already come within sight of it, though. In our normal life experience, we're closely confronted with increasing entropy as we age, suffer illness, and ultimately die. The cross symbolizes the various sorts of affliction of humans. And this symbol accomplishes something else that is decisive: it *interprets* suffering. Christians believe that enfolded within the darkness of the cross is the promise of Easter light. I'd like to envision a link between these older convictions and the cosmic necessity you described. In much the same way that a more complex structure must be purchased at the price of increased entropy, Easter's new creation requires the event at the cross.

SHELDON: Careful! In physics, structure formation is in no way a *necessary* result of increased entropy. And it is far more probable that open systems will disintegrate than evolve to a higher order.

THOMAS: That the origin of something new isn't mandatory fits my thinking very well. It wasn't at all inevitable that the cross should have led to the resurrection. Easter was a miracle precisely because it was quite unforeseen, something no one could have expected.

SHELDON: I feel my entropy increasing without enhancing mental order. It's time for us to get some sleep.

THOMAS: Yes, I keep forgetting. Our sixteen-day orbit around Saturn doesn't provide us with real day or night.

SHELDON: That's why we've maintained an earthly calendar and coordinated our clocks with Greenwich Mean Time.

THOMAS: I look forward to slipping into my hammock and turning off the light. Sleeping at zero gravity is better than normal sleep on Earth.

SHELDON: It is. Good night.

THOMAS: Good night. But before sleeping, I'll be making another entry in my diary about the cosmic dimensions of Easter.

A Hymn on the Cosmic Christ
From Thomas's Diary (April 4)

Today's conversation still echoes within me. What an incredible thought that the whole cosmos is witness to the death and life of Christ! I recall an early Christian hymn in which the Son of God is portrayed with the colors of Jewish "wisdom," the Sophia, that accompanied God's creational work from the beginning.[3] I have loaded the biblical hymn (Colossians 1:15–20)[4] from my private electronic library onto the monitor. The first stanza reads:

He is the image of the invisible God,
The firstborn of all creation,
For in him all things were created
In heaven and on earth,
Things visible and invisible,
Whether thrones or dominions,
Or rulers or powers;
All things have been created through him and for him
He himself is before all things,
And in him all things hold together.
He is the head of the body, the church.

The notion of Christ as the "image of God" opens two wide windows. The first window: God, the creator of this great universe, revealed his heart in the figure of Christ. More precisely, in the way Jesus died out of love for humankind.

We could not perceive what God is, had it not been manifested in the life and death of Jesus.

The second window: the entire creation takes part in this image, to which Christ was made. The beam of love extending from the invisible God to the visible Christ reaches deep into the vastness of the cosmos. Everything is filled with and carried by the divine Christ—the heavenly realm of the angels as well as the earthly empires of humankind with their rebellious rulers. Everything, even that which is furthest from God, reflects in its way the benevolence of the creator God revealed in Christ.

Greek prophets and philosophers spoke in awe of the cosmic Logos as Zeus, the god of heaven, whose body forms the whole universe: "Zeus is the first and Zeus is the last, . . . Zeus is the head, Zeus is the middle; from Zeus all things are made."[5]

Christians discovered Jesus Christ in that God of the universe. The singer of the above hymn describes Christ as "head" who has the church as his "body." As the Son of God penetrates the whole cosmos, he makes it correspond to his own living body. This alludes to the philosophers' pantheistic belief in God. All creatures are made and meant for praising and glorifying in their way God the creator—not only people all over the world but also angels and stars, elements, plants, and animals.[6] *They all cohere in* one *large cosmic community, in the "church."*

My view through the porthole becomes lost in the pitch-black depths of outer space. Myriads of distant stars gleam and sparkle. Their light delivers messages of unimaginable spaces and times, but it has nothing to proclaim to us today about the divine Creator and his unfathomable love. It breathes only coldness, isolation, and loneliness. The light of cosmic space becomes warm and welcoming only when a different light shines forth. When the light of the living Christ falls into the universe, the cosmos reflects it back a thousandfold.

The singer of the hymn knew nothing of the mighty spaces on whose coasts we camp here in Saturn's system. Yet in his time, he was not a backward hillbilly, but an expert authority in advanced philosophical cosmology. He understood the subtle play of prepositions—"in him," "through him," "for him," with which the metaphysicians of that era tried to describe the structure of the cosmos. He understood the universe as creation in the light of Christ. Even today, his insights are worth appropriating, though doing so in the light of contemporary science is more difficult than it once was.

Ice-cold Saturn again comes into my view through the porthole. Its moons Rhea and Dione now pass by silently as if in slow motion in front of Saturn's large globe. For billions of years they have been orbiting nearly uniformly, every few days, around their mother planet. Both moons are covered with a thick coating of ice. My vision of the cosmic Christ suddenly vanishes in the face of this titanic world of death. I turn back to the text that had lifted me up into bright heights.

The hymn rings out in a second stanza:

*He is the beginning,
The firstborn from the dead,
So that he might come to have first place in everything.*

For in him all the fullness of God was pleased to dwell,
And through him God was pleased to reconcile to himself all things,
By making peace through the blood of his cross,
Whether on earth or in heaven.

What had not previously risen into view—strife and war, suffering and death, blood and cross—now dominates the picture. In the first stanza, the cosmos reflects the divine magnificence of Christ in each of its particles. A heavenly perspective prevails; the singer looks with angel's eyes upon a world that seems to rest timelessly in God. The second stanza, however, directs our attention to a bloody drama enacted in the depths of the earth, to Good Friday and Easter. The cosmos is no longer ruled by harmony and symmetry but stricken by catastrophes and shocks. Now everything revolves around the single event on a remote planet, one episode in the long course of Earth's history. Jesus dies and is resurrected to life to reconcile to God what is torn in the universe. In the first stanza, the cosmic eternity is celebrated. In the second, a dramatic earthly story narrated that finally includes the whole cosmos.

What a colossal diptych! If one looks simultaneously at both pictures, the suffering and resurrected Christ becomes the basic pattern, according to which all of creation occurs. The Son of God does not descend only once into the depths of the earthly world but is always bearing anew suffering and resurrection. The pattern is rediscovered with each death, and at the origin of all new beginnings. His history is not a faraway incident of human experience but the pattern behind countless formations of order and new structures. What the narrative of Jesus sets forth as a unique event is the basic principle of cosmic evolution.

At the same time, this event suggests a fullness of life that will totally penetrate and change creation in the future. Thus Jesus Christ is the prototype of *the present* as well as *the coming eons:* "I am the Alpha and the Omega, the first and the last, the beginning and the end."[7]

The view out of my porthole lets me trace the vast dimensions that the Creator has spread out. In my thoughts, I have left historical Palestine as *far behind as we have left Earth behind us in the distance. The incredible vastness of space and time has swept away the floor from under my feet. How insignificant do the narratives from Galilee and Jerusalem seem to me here—and how I long to find myself there again. My soul is like a tree: its crown stretches across wide spaces, but its roots seek firm footing and support in the earth.*

Despite my attraction to thinking about cosmic dimensions of the crucified and resurrected Christ, I cannot forget that the cross was a gallows and an instrument of torture that destroyed the life of a lonely and weak human being. This recollection protects me from identifying the ordeal of Jesus with the turbulences of physical processes. His story casts light on the restless movements of the universe, but it cannot be absorbed into a timeless cosmological symbolism. Sciences foster a different understanding of knowing. From a scientific perspective, one need not tell stories to meet the truth. The mathematical formulae apply independently of one's time or place. Can I make it clear to Sheldon why my old hymn cannot be satisfied with the first stanza?

Creation in Different Worldviews

SHELDON: Good morning, Thomas. It's time for me to contact PAN, the unmanned Penetrating Automatic Navigator traversing Saturn's ring system.

THOMAS: All right, Sheldon. While you're doing that, I'll tend to the bio-cultures for fresh food. According to our Earth calendar, it's been three years since the *Hermes* mission left Earth's orbit. So I'm proposing our special celebrative menu will be nori (edible seaweed) garnished with liverwort, bean algae, and soft-boiled symbiotic lichen mushrooms.

SHELDON: Sounds odd but no doubt you'll make it tasty.

THOMAS: I appreciate your confidence and wish you felt the same way about my theology.

SHELDON: Sorry about that, my friend, but I'm still struggling with our recent dialogue. You mentioned a while ago that some of the philosophers of antiquity imagined the universe as a huge organism, animated by a divine spirit. In those days, it was common to think poetically. Why do you turn now to modern scientific concepts that are characterized by a high degree of abstraction? The traditional symbols were nourished with phenomena directly perceived in everyday life. This immediacy has nothing to do with today's scientific analysis, which looks behind immediate impressions and feelings and which even distrusts them.

THOMAS: I don't agree. Cosmology was an integrative discipline for the ancients, combining careful observations of nature with philosophical reflection and religious ideas. The biblical texts and early Christian theologians integrated generally accepted cosmological worldviews of their time into their messages. They didn't proceed solely on the basis of superficial appearances and impressions. The well-known story of creation at the beginning of the Old Testament, for instance, makes theological use of ancient Babylonian accounts of the world's creation. God, Satan, Adam, and Eve have counterparts in this genesis story written centuries before the Judeo-Christian Old Testament.

SHELDON: Doesn't that put you in perilous dependence on the worldviews that are currently in fashion? Because theologians had shackled themselves to the geocentric cosmos of Ptolemy in the Middle Ages, they had great difficulty in accepting the Copernican revolution, which proved that the sun, not Earth, is the center of our solar system and our Earth is one of the planets revolving around it.

THOMAS: We can't blame theology for operating in accord with worldviews of a given era. Theology has to be conducted within the context of the worldview that prevails during the time of a theologian. That's why it interpreted the Babylonian and later the Greek worldviews autonomously. Theologians incorporated their experience of God quite deliberately into the cosmology then prevailing. In doing so, the universe got a new face throughout the history of civilization.

SHELDON: How did that work over the centuries?

THOMAS: Clearly, to be sure. The Babylonians declared the sun, moon, and stars as gods. In the biblical account of creation, they were downgraded to sources of light. In return, time gains a major role in the first chapter of the Bible. Centuries later, Christian theologians disputed the eternity of the world, as previously set forth in Platonic and Aristotelian cosmology. In their view, God actually created the world out of nothing.

"Well, my friends," Randall sighed. "That's over my head. But it's important, because the biblical concepts of creation have played a paramount role for a long time in the dispute between natural science and theology. Yesterday I sent a message to Professor Claudia Estermann from Heidelberg, asking her to give us her most recent thoughts on foundational knowledge drawn from present-day interpretations."

A Six-Day Workshop Report

The first creation narrative of the Bible (Genesis 1:1–2:4a) originates from the sixth century BC, therefore from the epoch of the Babylonian exile. It is the distinctly systematic and well-wrought prelude of an extensive theological work of priestly origin. It should be distinguished from the probably older second creation narration (Genesis 2:4b–25), which mainly deals with the creation of man and belongs in the larger context of the story about the paradise (Genesis 2:4b–3:24). In the priestly account, the living spaces (heaven, sea, air, and earth) are created in the first three days. In the following three days, the creatures inhabiting these areas are assigned. The timing is in prime focus: The first creation work, the light, makes the sequence of the creation days possible. The celestial bodies created on the fourth day regulate days, weeks, and years. Contrary to a widespread misunderstanding, man is not the crown of creation: creation is completed only with the seventh day, whereupon Israel answers with its Sabbath practice and its cult.

Haubensak is correct in his statement that Genesis 1, to a large extent, works with widespread ideas of ancient oriental cosmologies, particularly the Babylonian. However, because of his cosmological obsession, he tends to overestimate the continuity between the ancient Babylonian and biblical worldview. The priestly source's interests are exclusively theological, not cosmological. For example, the creator God calls things into being in such sovereign fashion that the primeval chaotic matter almost evaporates.

It should be especially pointed out that texts of this kind are intended not to clarify speculative questions regarding life's origins but rather to address contemporary problems of their time. Creation is presented in Genesis 1 as an orderly abode for life, built on a reliable foundation and into whose spaces humans, together with all other creatures, are admitted. Therefore, modern physical cosmologies and the biblical creation stories should not be viewed as competing with each other, since they try to answer totally different questions.

From a critical perspective, I must comment on Haubensak's excessively restricted outlook concerning creation. He concentrates so heavily on Easter as a new creation that he misses the full breadth of the Old Testament's creation traditions. He tends to slight the Hebrew Bible's emphasis on the reliability of life that God grants. God not only creates new things but also bestows fertility and abundance. Perhaps this deficit has to do with the artificial environment of Haubensak's theological discussions with the physicist Sheldon Cutter and his nightly diary entries and meditations.

SHELDON: So, do you agree with what I have heard theologians saying? You accuse science to having blind spots?

THOMAS: Yes, the light upon which religion is based shines also into the twilight zones of science as well as the knowledge system operative in a given time. Religious experiences can dare to witness what the zeitgeist, or spirit of the age, tries to suppress. Theology shares this critical function with art and with philosophy.[1]

SHELDON: So you really believe that science overlooks a significant part of reality?

THOMAS: You bet! You're downright specialists in the art of perceiving reality in such a way that it must fall silent.

SHELDON: Can you give me just one concrete example of that?

THOMAS: You totally ignore suffering. From the viewpoint of methodology, there is no difference between a physical experiment and the vivisection of an animal. In addition, there's a problem we talked about earlier: the result of the cognitive process, the gained knowledge, doesn't express what price was paid. As fascinating as the progress in neurophysiology has been, I still can't forget that this knowledge was bought at the price of anonymous suffering by countless animals.

SHELDON: Yes . . . I can see your point. Many scientific experiments are based on destructive activities, especially when humans themselves become research and discovery test objects. But your God, the Creator, also seems to have incorporated destruction and suffering into the organization of his world. Planets can only be formed out of the ashes of former stars. We humans and our civilization can only develop because billions of animals and plants were barbarously outsmarted, killed, and devoured.

THOMAS: You've touched upon a deep problem in theology—namely, why God allows evil in his creation. Almost all answers to this question are at risk of sounding cheap, because abstractions seem to deny the concrete actuality of suffering. I'll even admit that science has sensitized us to the dreadful stress that accompanies the evolution of species and the fate of individual creatures. Catastrophes, competition, and struggles for living space often characterize the beginning of new forms of life. The ancient Greek authority Heraclitus was right in the end: "War is the father and king of all."[2] This tragic perception finds its resonance in the Easter story when Jesus suffers and, with him, God himself. Here, another aspect of God is revealed, distinct from the image of the One high above all enthroned rulers of the world, who looks down unaffected by the wrongdoing of his creatures. The Son of God shares the suffering of all creatures, even taking it upon himself in their place.

SHELDON: I can see the parallel you're trying to make and agree that human suffering is a universal phenomenon that both science and religion have dealt with in their way, though with less-than-perfect results.

But now that I've received news from PAN, which successfully traversed the ring yesterday, let me report the results in a "Postcard from Saturn for the Cyber School." Here's what I've come up with:

From Darkness to Light
A Postcard from Saturn for the Cyber School

Hello, Earth!
The Saturn expedition sent out the robot probe PAN *(the Penetrating Automatic Navigator) to the ring system on March 20, 2051. It works automatically for the most part and reports its observations directly to Earth. Here in the* Hermes Trismegistus, *we only took over the supervision when crossing the B ring,[3] since radio transmission from Saturn to Earth and back takes about three hours.*

PAN approached the ring from the side opposite to the sun. In the ring's shadow, it wasn't easy to recognize the first rocks and boulders in our large-screen projection on the cabin wall. But PAN's onboard radar kept the distances under control; the probe slowed down automatically. Then it dove into the ring like into a downy pillow. The ring particles floated like dirty snowflakes in space and hid the sunlight. They bounced off the protective shield of the slowly moving probe. Every so often, larger rocks slid into view and pitch-black lumps of ice came spookily toward us. When one floated in the way, it was gently pushed to the side by the probe's intelligent gripper arms and, at the same time, examined by its sensors. Thus the probe automatically groped its way through the ring, which is less than a hundred yards thick.

Our job was to recognize large obstacles that the probe could not clear away and to navigate PAN alongside them. We followed the crossing via the large-screen projection with highest concentration, like a car driver in a snowstorm. The ring was extremely dense on one stretch of thirty yards. Large rocks popped up and the interspaces were filled with debris from past collisions. Colossal giants suddenly emerged several times out of the darkness. We were able to get them in sight with PAN's remote-controlled camera and maneuver around them.

After three hours of tense focus and navigation, it became brighter. The sun began to appear, dull at first in the increasingly thinning dust but giving us a better orientation. A ring-shaped play of colors began to sparkle around the center of the light spot. It must have been triggered by the refraction of light and its reflection on freshly broken ice crystals and small ice particles. They draped the ring like a lightweight down. A flickering reddish circle surrounded by bluish light encased the defuse light of the sun. The sight was like a fringed, blurred rainbow with reversed color sequence. Snow crystals twinkled, and fractured areas of ice lumps began to glimmer bluish.

Then suddenly we came out into the open, above the Saturn ring. We had arrived on the other side of the disk of ice, dust, and threatening rocks, in the midst of glaring sunlight. A clear expanse suddenly opened up, as when an airplane comes out of a sea of fog. It became impossible for the camera to transmit the contrast realistically. The sun was dazzling in the pitch-black sky. The B ring expanded behind us, blindingly white and seemingly infinite. It gleams and glitters where the sun is reflected. In the far distance, a curved black band interrupted the plain, through which stars sparkle: the Cassini division. Behind that lies the A ring, a still unexplored luminous world.

We can pan the camera to the opposite side and see the gigantic planet Saturn. It appears white, bright, and wide, like a perfect heavenly sphere. Its bright cloud cover shows only some weak yellowish structures that disclose its rotation. It turns majestically around its axis like a weightless spinning top hanging in the ring.

From Saturn with best regards, Sheldon Cutter

Different Perceptions in Different Perspectives

SHELDON: OK. Do you think this is the kind of public report they want for their Cyber School program?

THOMAS: Yes, it's fine, Sheldon. Good work. It combines scientific objectivity with personal experience. That brings me back to a point I've been wanting to make. I have the impression that the type of perception we adopt in science and religion is fundamentally different. This may be the key to many misunderstandings. We don't seem to be talking about the same thing, even though it sounds similar.

SHELDON: That may very well be true. I remember when I heard undergraduate Darwinists and Creationists argue with each other at North Dakota State University, I had the feeling that the modern understanding of evolution had a perspective that was totally different than the biblical account of creation. Thus the controversy was absurd from the beginning.

THOMAS: Indeed. That's because science tries to see the world from the outside, while religion deals with a special kind of internal perspective.[1]

SHELDON: Not at all. It's because research seeks the greatest possible objectivity, whereas religion deals with a subjective affair, with which some agree and others don't.

THOMAS: That's not what I meant! I don't like the idea of separating "subjective" from "objective." You're expressing an attitude that unfortunately has long been socially accepted—that religion has become a private matter, that it no longer contributes to find the truth in any generally accepted way, and that religion has become arbitrary.

SHELDON: Yes, that's what I'm saying.

THOMAS: And you're wrong. Religion doesn't conform to the model of subjective and objective. It appeals to people directly, before any contrived duality.

SHELDON: But you yourself mentioned an inner perspective.

THOMAS: By inner perspective, I mean a perception where a person takes part in experiencing reality. This kind of participatory perspective can never be gained from the outside because it's based on the relationship between the part and the whole. In religious language, this whole is expressed by the notion of "heaven and earth."

SHELDON: By the "whole," do you refer to God himself?

THOMAS: There are various religious views on this point. Heaven and earth is more than the universe and includes the human beings and their mind, all that can be regarded as the work of the transcendent creator God. In any case, the whole has to do with a fundamental reality that is experienced as unknowable and to which humankind owes its existence. To partake in this whole is felt as a gift. You can't experience it from a distance. The method of modern science is totally different. Science can't see the world in this way.

SHELDON: You seem to be looking for the blind spot in science's point of view. But the fact is that location doesn't play any role at all in science. I'm sure that intelligent beings in other planetary systems, if they exist, measure the same speed of light and find the same benzene ring. But let's get back to religion again. I don't quite understand what you mean by religious perception. Evidently for you, it has to do with more than just certain views about the world and life that inevitably must stay subjective.

THOMAS: Religious traditions have developed a very special way of perceiving the world. In this kind of awareness, the world becomes transparent toward its creative origin. A "third eye" is opened, and we look with these new eyes into the world that reveals itself in a light that hasn't shone before. The religions talk about inspiration or revelation, and they express this perception of the world as a gift in different linguistic forms.

SHELDON: Your description reminds me of the development of three-dimensional imaging technology. During the past two decades, the visualization of three-dimensional space was developed in multimedia. Could you say that through your "third eye" the usual sensation is enlarged by another special dimension, similar to how a two-dimensional picture can suddenly gain depth?

THOMAS: Yes, it has something to do with sensing a kind of depth. I found an old picture book in the attic when I was a kid, where the illustrations became three-dimensional when my eyes properly focused on them. I could literally observe how a new dimension began to build up. And I've compared religious perception to this ever since. The world gains in height and depth. This perception of a depth dimension comes into play especially in a mystic ecstasy.

SHELDON: Then I only wonder why we don't register this reality with our scientific methods. Why can't what is perceived in religion be identified and observed due to some emission mechanism of photons?

THOMAS: No, no. The detected contents of the photons can't be measured by mathematically formulated laws. The visionary owes his seeing to the "eye of his soul," a special receptive organ.

SHELDON: That "eye of his soul" takes us again back to psychic conditions with an altered state of consciousness. For us scientists, the matter-of-fact, rational consciousness is the basis of all research work. There's no room for mystic inspirations.

THOMAS: The point of religion is to practice the perspectives that open up in the daily context, like mystical experiences on a mountain peak or being swept away by a work of musical art, such as *St. Matthew Passion*. I imagine that in strict scientific work, inspiration, lightning-fast insights, can give you the same thrill and value.

SHELDON: Inspirations are certainly important in science, but the crucial point is what comes after from critical, empirical, or theoretical analysis that can be reexamined and reexperienced by anyone. It's different with religion. I can agree with you that the opening of a "third eye" may yield another kind of sensation of the world. But maybe you're not dealing with reality but rather experiencing a delusion, or a "wishful-seeing." But just a moment, we're getting a signal from PAN.

THOMAS: Better see where it is in the ring system.

SHELDON: Probably have to take over manual steering again, to avoid collisions with colossal ring particles and ice rocks.

THOMAS: I have confidence in your ability to keep us safe, Sheldon. And while you're taking care of this, I will go to check my mailbox. I have asked the Study Unit for the Perception of Reality (SUPR) of the University of Zurich, Switzerland, for an expert's opinion regarding the interdisciplinarity dialogue in which you and I are engaged. Professors of natural science as well as the humanities and human sciences participate in the group.

Critical Assessment of Interdisciplinary Dialogue

Simply stated, at least four models for the relationship between natural science and theology (resp. Christian faith) can be identified: *conflict*, *independence*, *dialogue*, and *integration*.[2] Although all four approaches are well observed in the discussions from the world of Saturn, the dialogue obviously enjoys highest priority.

Our study group takes note with satisfaction that the physicist Sheldon Cutter and the theologian Thomas Haubensak do not drift off to an all-embracing super theory but rather cultivate a culture of *open conversation*. The individual discussions, therefore, do not unfold their subject systematically but instead circle around and deepen the theme in repeated approaches. The discussion creates a space for *interactions* between the particular perspectives, involving the presentation or formulation of the problems resulting in expressions of doubt and uncertainty, and not only for rhetorical effect, plus straightforward conflicts, and convergences. We notice that the two opponents do not come to a full mutual understanding but instead remain with their different positions—positions that are generally not contrary but are to be characterized as *contactless*. This is no coincidence. In our opinion, the dialogue's main difficulty between two very different disciplines is that the statements are placed on totally different planes that are very difficult to mediate. This inhibits real contact, but it does not lead to clear antitheses.

With this, we have already touched the *limits* of the bridges proposed between these two cultures. Four items are especially noteworthy:

1. Perhaps most noticeable is the absence of more demanding epistemological perspectives. The course of conversation between our Saturn travelers has something nonchalant about it that now and then evokes the danger of *narrowing the horizon* and of *short-circuiting* things. In our opinion, this deficit in perspective is not entirely negative. The complicated reflections on the theory of science and analytical philosophy often lead the dialogue between natural science and theology into mutual doubt and puzzlement. When the problems pile up to the sky, most partners lose interest in talking, and after all their effort, nothing really results. On the other hand, the blind spots that the astronauts have because of their specific education and interests facilitate many noteworthy similarities.
2. With some uneasiness, a number of members of our study group noticed the marginal placement of *ethical questions*, particularly in view of the pressing problems of biotechnology and reproductive medicine. The contemplative situation of speakers in the outer region of the solar system, far from all earthly day-to-day events, may be the reason. Both partners seem largely to agree that ethical orientations are based on an ontological ground, that the question about the *good* cannot dispense with the question of

what is *true*. The majority of our study group feels that this position is outdated, since it is incompatible with the circumstances of modern and pluralistic ethical discourses.
3. The *methodological problem* requires deepened reflections. The sciences working empirically and mathematically have moved far away from the hermeneutically oriented philosophy and theology. In fact, both astronauts circled several times around this question, but did not come further than the traditional definition of contrast, where science has the task "to explain" and the humanities "to understand." Considerations regarding the particular grammar of the very different specific languages of both disciplines could lead further.
4. The theological members of our group criticized sharply how Haubensak interprets Easter as the basic pattern for the "central order of the world" that evolves stepwise through the formation of transitory structures. Easter, however, is a unique event that changed forever the course of time. It is the antithesis to the paradigm of evolution.

SHELDON: Thanks, Thomas. I appreciate your confidence, and I was in fact able to steer the PAN probe through the rings and safely out the other side. But there's another job I have to do now, which is report on our much-traveled ODYSSEUS probing the moons of the Saturn system. It's a great opportunity for another one of my postcards for the Cyber School program.

Figure 4. The mysterious Saturn moon Iapetus always shows the same side to its planet. In the orbital forward direction (left), it is pitch-dark and reflects only a few percent of the sunlight. Its average density is 1.2 times that of water. Iapetus probably contains an immense amount of water ice. (Photo: NASA/JPL-Caltech/Space Science Institute)

A Monster Moves
A Postcard from Saturn for the Cyber School

Hello, Earth!
Today I'd like to report about the unmanned probe ODYSSEUS *(Orbiting Device for Yearlong Saturnian System Exploration and Unrestrained Studies). Monitoring* ODYSSEUS *is one of the astronauts' most important activities. It's also their job to decide when and how to intervene at critical moments without delay.*

The tension and vigilance have risen every time ODYSSEUS'S *cameras came in close proximity to the bizarre and violent worlds outside our cabin. We often have had to change its course. Fortunately, the probe has enough fuel and a wide variety of instruments to observe every object in Saturn's realm, especially the moons circling around it, and making detailed observations.*

With complicated orbital maneuvers and the help of the moons' gravity, ODYSSEUS *swings through Saturn's empire and visits most of its 82 moons and over 150 moonlets embedded in the rings. The largest moons had already been detected in the seventeenth century and bear names from the ancient Titans, a powerful pre-Olympic dynasty of deities, the children and grandchildren of Uranus and Gaia. A long-forgotten mythic past comes alive before our eyes.*

When the probe crossed the orbit of Iapetus in March, the mass spectrometer detected an enhanced gas density. That's why we're steering ODYSSEUS, *with several swing-bys on Titan and Rhea, into an orbit around the strange moon.*

Iapetus is one of the most mysterious and unknown objects here in the outer region of Saturn's system. Despite its similar size with Rhea, it gives the impression of a stranger among the companions of Saturn. With a density of 1.2 times that of water, Iapetus must contain still other components besides ice. We don't know yet what they are, but it may have a core composed of very heavy elements.

Iapetus's most evident mark is a jet-black region that covers almost half of its surface.[3] *It's so dark that one cannot see any structures from*

Earth, even with the best telescopes. Since Iapetus revolves around itself once per orbit around Saturn, similar to Earth's moon, the dark region always faces in the same direction—that is, forward. The ancient god always shows a grim visage. Is it only a mask? What's concealed behind its face of death?

While we still brood over the moon's cryptic secret, the submerged titanic primeval world breaks tumultuously into our present era. Iapetus surprised us on April 13 with a violent volcano eruption in the middle of the dark area. On the sensational pictures from ODYSSEUS, *which we beam onto our cabin wall, it behaves like a primeval monster. The volcano doesn't hurl out red-hot lava but a black powderlike liquid. This wallows around like a sluggish cloud in slow motion, rising up a quarter of the moon's radius and sinking back down to the moon in seething movements due to the low gravity. From there it streams along the ground from the volcano in dark ghostly waves. Finally, Iapetus has given us a glimpse into the bubbling life behind his death mask!*

The optical spectrometer discovered not only liquid water on Iapetus but also methane and other organic compounds.[4] *The spectral bands in the infrared range resemble those of tar particles. We assume that water reacts with methane and is transformed to acetylene and hydrogen cyanide under the influence of the solar irradiation. The hydrocarbons condense to tar-like frozen particles that are carried along with the volcanic eruption. While still in flight, the water must solidify in Iapetus's icy coldness.*

With a surface temperature of -350°F, the liquid water on this moon is a marvel. There must be an enigmatic source of energy in the core.

But Iapetus has also a light countenance. On the backward-facing surface, one can recognize areas that glisten magnificently in the sunlight. These spots are probably hundreds of miles of thick layers of ice of great purity.

I'm overcome by a fierce thirst at the thought of this precious commodity in our artificial world, where we must content ourselves with stale water that's been recycled a hundred times. And I'd love to race with a jet sled on the wide ice sheets as I did in my youth in the North Dakota plains!

We eagerly await our second manned excursion that will bring us closer to this large moon. Which face will the mysterious Iapetus show us?

From Saturn with best regards, Sheldon Cutter

The Language of Images

SHELDON: Hope this second weekly report for the Cyber School program is pictorial enough. Now, during our lunch break, we can download into the parallel computer the latest images from the ARGOS, the hundred-foot telescope on the mother ship that was used during an especially favorable window between the sun, Saturn, and Titan for the observation of the Uranian system.

THOMAS: What can you tell me about Uranus?

SHELDON: The bluish Uranus, named after the Greek sky god, is, from Saturn, the next planet farther out in the solar system. It was discovered only in 1781.

THOMAS: Can we take a look at what ARGOS was able to see?

SHELDON: Certainly. Let me project them at the cabin wall. You will have to wait some seconds to let them gradually emerge and sharpen by the transformation algorithm.

THOMAS: OK.

SHELDON: These pictures reveal the strange world of little-known Uranus, which is, like Jupiter and Saturn, surrounded by numerous moons and rings. Especially impressive are the new pictures of the moon Miranda, which must have suffered a massive grazing collision in its early period. At the beginning, there are only totally amorphous volumes of data; then, thanks to gigantic computer programs, the pictures get closer and closer to reality.

THOMAS: I'm fascinated at how we can produce pictures with the help of a sophisticated, ingenious mathematical transformation. The method seems to be analogous to interpreting religious perceptions.

SHELDON: It's not the first time I've posed this question: What do these analogies really mean to you? They remind me of your attempts to forge religious parables out of our cold scientific facts.

THOMAS: I interpret my perception of reality with patterns. I start with religious experiences of crisis and renewal, and make the surprising discovery that this pattern often recurs in events outside of human life in the vastness of the universe.

SHELDON: What do you mean by pattern recognition? Recognition of patterns plays an enormous role in robotic and computer science. The accuracy with which our brain recognizes a known human face is another example. Also, temporal sequences can form a pattern.

THOMAS: Now you're getting too scientific for me to understand.

SHELDON: It's not easy to explain or understand, not even for scientists, but I'll try.

THOMAS: Thank you.

SHELDON: Here is an example from wave theory: When free energy is present, small fluctuations can increase by feedback and become unstable waves. They grow until the energy reservoir is depleted or a nonlinear follow-up process changes the conditions. If the instability gets saturated in one of these ways, the waves remain stationary and finally disintegrate again. This pattern can be observed in many different types of waves. To recognize a pattern, the model and the sample must be of the same kind. But this is not at all the case with your Easter interpretation of evolutionary processes. There are partial agreements but also striking differences. Above all, they are on totally different planes of experience. You're mixing up different terms and talking like a poet.

THOMAS: Your impression is correct. When I search for analogies between the religious world of experience and the cosmic reality, I cross the stringent frame of concepts. I take advantage of the metaphoric power of language.[1]

SHELDON: This kind of language is inaccurate, subjective, and, for me, the epitome of unscientific thinking.

THOMAS: But wait ... listen ... In formal rhetoric, metaphor is a linguistic form that compares two essentially dissimilar things. "I'm in seventh heaven" is a beautiful example of a good metaphor. It would be ridiculous to think that a person in love would literally float around somewhere up in space. And yet this sentence fittingly expresses the lofty feeling of happiness. Only for us astronauts would this statement have a halfway realistic meaning, particularly here in the sphere of Saturn, the seventh planet of antique astronomy.

SHELDON: So, metaphors work like mathematical equations that relate two or more physical parameters to each other?

THOMAS: No, metaphors are not formulas. A metaphor dwells on the richness of our daily experience. If it works, it expresses the unknown in a way that conceptual language could never provide. The metaphor accomplishes something like a small act of creation; it pulls us into its linguistic play and illuminates something we couldn't describe otherwise. That's why metaphors have such importance in poetry and religion.

SHELDON: What does that mean, then, for Easter? What's the image and what's the thing it stands for?

THOMAS: Recently, the spontaneous emergence of something new in Saturn's atmosphere—an ice storm, a blinding white hurricane of frozen methane—filled us with awe and enthusiasm. I interpret that as a metaphor for the divine creation of something anew. But I must admit, from the earthly perspective, the analogy between Easter and a developing tempest is out of place—a not very successful metaphor! Nevertheless, being up here, out in space above a planet where no life is harmed, inspires me to such a bridging of the two planes.

SHELDON: Which gap do you actually want to bridge? You not only transferred the pattern "emergence of new from old" to Good Friday and Easter, but you also reversed the "Easter paradigm" to interpret cosmic processes.

THOMAS: You're a keen observer. Actually, I tried both: to shed light on Easter from cosmic phenomena *and* the opposite, to see the cosmos in the light of Easter. What develops spontaneously in the universe then becomes a reflection of divine creation.

SHELDON: It seems to me that your metaphor is like an equation with unknowns on both sides. You connect personal experiences on one side with theological convictions like Easter on the other.

THOMAS: I try to portray the inexpressibly divine with images from the familiar world so that the metaphor flows from the world to God. This can result in a feedback effect: with the metaphoric connection, a new light is cast *back* onto the familiar world.

SHELDON: Sounds vague and dubious to me.

THOMAS: Let's look at the common religious metaphor "God is light."[2] The unknown, God, is illustrated here with something familiar—the light. Yet God isn't identified with physical light. The image preserves God's invisibility; however, it brings to mind all those qualities that constitute his presence: brightness, warmth, orientation, perception, and beauty. Light as such is the epitome of vitality. At the same time, it can blind and even kill in its overabundance. This is also part of God's nature.

SHELDON: I certainly agree at this very moment that light is something special. Imagine we were lying on a beach, sunbathing in a flood of Caribbean sun. The sun is so weak and small out here, in the coldness of the universe.

THOMAS: Yes, I share your desire for the sun. Warm, life-friendly sunlight is a delightful gift. On Earth, we take it far too much for granted. Without sunlight our soul becomes stunted and wastes away. Light reminds me of the mystery of Easter and becomes a metaphor for it: a wonderful spring day after a cold winter, the birth of a child, the sprouting of a plant out of the ground. The metaphor sheds light both ways: Easter is like the great springtime, and the pleasant spring day or the birth of a child is like the

Easter light. Metaphors like this are building blocks for an encompassing theology of creation.

SHELDON: These images have a direct appeal to me. But they have nothing to do with the reality I confront in my science. We live in the year 2051 and move in an artificial biosphere of a complicated machine through the icy moon system of Saturn. All images from our familiar earthly surroundings lose their evident nature here. You talked about God as light. For me as a physicist, the laws of electromagnetic radiation have nothing to do that would be remotely comparable with God.

THOMAS: You're asking whether the old metaphors have lost their power in the context of present-day life and thought. This is exactly what's driving my buccaneer raids into science. What does it mean when we try to link religion's metaphoric language with the descriptive, objectifying language of physical equations and formulas? Here we're no longer in the area of daily language and its familiar images.

SHELDON: So now you've finally revealed the hidden agenda you're trying to pursue onboard this spacecraft! The victory of mystical religion over empirical science.

THOMAS: No, not exactly. More closing the gap. For many of our contemporaries on Earth, there's an increasingly large gap between the worlds of religious experience and scientific knowledge. But the cosmic perspective of our *Hermes* mission convinces me of the opposite. The chaotic emergence of new entities discovered by modern science even deepens my understanding of Easter.

SHELDON: I respect and appreciate that being out here in space with me helps you to reduce the gap between religion and science. So, let's take another look at the earthly sphere. I'll try to see things from your perspective and figure it like this: Easter is like the emergence of life on the early Earth, which had been bombarded by asteroids and comets, and devastated by volcanoes.[3]

THOMAS: Good! That's a beautiful metaphor because the image of Earth, still young in the midst of catastrophic events, appeals to us directly. It draws on our experience of spring and the birth of a child. Metaphors become more difficult when we leave this type of fundamental human experience behind and try to use complex processes in physics and chemistry that are difficult to comprehend.

SHELDON: But remember the mathematically reconstructed pictures of ARGOS and the buildup of cyclones and tornadoes. The power of metaphors breaks down in the realm of science. At best it only has a didactic value.

THOMAS: You distinguish too sharply between everyday experience and science. Our mundane worlds of experience are also determined by modern knowledge. Copernicus, Darwin, Einstein, and Planck fundamentally changed our lives. Your metaphor of the young Earth remains incomprehensible without scientific knowledge.

SHELDON: I won't deny that there are transitions between ordinary and scientific experience of the world. I liked to use images and allegories in popular science lectures to make abstract physical equations understandable. However, I was amazed how the audience transferred these images into everyday situations and enriched them with their own experiences. When I mention that massive stars have a much shorter lifetime than low-mass stars, I can sense how overweight people get uneasy. When I describe how old stars explode and spread their accumulated heavy elements into space to be incorporated into the next generation of stars, parents identify with that, too. Since then, I have become more and more suspicious of imagery.

THOMAS: You don't live only in the sphere of mathematical formulas either, Sheldon. To get your bearing in life, you use interpretive patterns that draw upon images and comparisons. As rationally thinking beings, we can't exist without the help of such patterns in order to locate and orient ourselves. The problem isn't that you translate your abstract equations into worldviews and philosophies of life but that you don't reflect on this translation and therefore don't handle it critically.

SHELDON: Evolution certainly promoted this kind of orientation. Whether interpretative patterns contain any truth is a different question. I disagree, however, that I unwittingly make physics into a religion. It was science that makes it possible to do without religious worldviews, whereas you are still entangled in the web of old myths. Your language games are nothing but desperate efforts to feel a little more comfortable and cozier in a chaotic cold universe, where we humans constitute only a brief episode.

THOMAS: Your image of the universe as a hostile, icy desert threatening our existence is just as mythological as that of a protected garden or a living organism.

SHELDON: My suspicion of metaphorical language is renewed once more with this statement. You return to a mythological cosmology that decorates itself with scientific clichés. But . . . wait . . .

THOMAS: What is it, Sheldon?

SHELDON: We'll have to continue this later. ORPHEUS just contacted us from its orbit around Titan.

THOMAS: What's happening? Something wrong with the ORPHEUS?

SHELDON: Hold on . . . The Titanauts are having more problems with mapping Titan. Evidently, the data received from the radar measurements can't be processed for consistent pictures. Instead of improving, the pictures are becoming more blurred and ambiguous.

THOMAS: That could be dangerous for our fellow astronauts who are about to attempt a landing on Titan.

SHELDON: Without a doubt. I'd better start troubleshooting.

THOMAS: Off you go, my friend. I'll keep track of everything onboard here while I write in my diary.

Astrology—a Contentious Point
From Thomas's Diary (April 13)

During work in EDEN *this afternoon, I asked Sheldon how he felt about astrology. His reaction was exceptionally gruff. Visibly annoyed, he rejected my fainthearted suggestion that cosmic fields might be influenced by the movements of planets and could have an effect not to be ignored on Earth's organic life.*

"Every box and every building in the near vicinity," he responded, "has a substantially larger gravitational effect on us than celestial bodies in the far distance." For me, the casting of horoscopes soon became suspect after a youthful, enthusiastic phase driven by astrological software. Still, I don't want to throw astrology onto the trash heap of intellectual history. Its lasting value lies not in its prophecy of the future but in its differentiated psychological typology, which draws on a rich mythological legacy.

In discussing this topic, my godfather used to quote the witty words of the great German philosopher of nature Carl Friedrich von Weizsäcker: "Astrology is such an old science that it cannot be totally wrong."[4] *The names of the planets preserve remembrance of the old gods, the powers and forces between heaven and Earth. Our enlightened reason has declared them to be figments of imagination. But haven't they reappeared even more powerfully in the shadow cone of our restless flickering light of knowledge during the last hundred years?*

"Poppycock," Sheldon responded. "With all respect to you and your father, the imaginary configurations of stars and planets have no influence whatsoever on our character or relationships." He is probably right. But my father believed it and ... we've truly become children of Saturn, Sheldon, like hermits! In speaking of ancient mysterious beliefs, I can't get the ninth Tarot card with the hermit out of my mind: old, lonely, and abandoned.

"No, no, no!" Hoihong strongly objected here. "We just delete this astrology nonsense."

"But it is part of Thomas's intellectual world," Astraia responded. "Just the facts, but all of them . . ."

"Let's leave it in, my friends," Randall decided. "It's a good transition to what follows—life in the incredible reality of the quantum world."

Quantum Mechanics and Reality

On April 16, 2051, at 11:52 UT, the *Hermes Trismegistus*'s radio signals suddenly break off. The failure seems to be in the main transmitter of the spaceship, an optical transmission unit with a state-of-the-art quantum well laser.

Hoihong suspected that the power diode was no longer pumping.

"I am sure that IASA had selected an extremely reliable model and the best semiconductor from a large number. The real blunder, therefore, must have occurred when Sheldon attempted to replace the diode with the spare. During his work, he probably charged himself electrically and destroyed the replacement diode by touching it."

"Hmmm . . ." Randall said. "IASA likes to justify manned spaceflight by pointing out the irreplaceable role of human competence in making repairs. No wonder that Harold has kept so quiet about this."

"But the astronauts on the *Hermes Trismegistus* were able to start up the emergency transmitter," Astraia noted. "It took them a while because they had little instruction in this and had to work according to manuals."

Since communication with Earth was impeded, life on the spaceship becomes more monotonous than before. The euphoria that accompanied them when they arrived at Saturn's system and observed the storm is now gone, replaced with a sober, disenchanted mood. The defective transmitter dominates the theme during the first lunch after *Hermes Trismegistus* fell silent, but the dialogue went on.

THOMAS: I don't understand why the laser stopped working in the first place. Nothing moves and wears out in a diode, and in the vacuum of space there's no rust. I wonder if cosmic rays had an effect.

SHELDON: That's what I suspect as well. A high-energy hit can cause defects in the lattice structure of the atoms. When they're close to the active layer, the current through the quantum well is reduced, leading to failure of the laser.

THOMAS: What's a quantum well?

SHELDON: It's that layer of the diode that, according to classical physics, electrons can't cross. Thanks to the tunnel effect, however, some of them still make it through. At the same time, they trigger a light avalanche. Since it is a quantum mechanical effect, one can't picture the electrons as small particles. They're more like small diffuse clouds or wave packets. This is also only an image, of course, behind which the uncertainty of the quantum world hides.

At the risk of presenting you with something well known, Thomas, I suggest you read this passage about uncertainty from the *New Encyclopedia of Physics* (2048 edition).

Beyond Newton

In 1687, Isaac Newton simplified material bodies into points in space. His *Philosophiae naturalis principia mathematica* describes the movement of these matter points under the influence of forces. Newton's classical mechanics was replaced in the 1920s by quantum mechanics. In the new mechanics, these bodies could still be points, but their position is not exactly known. They are only given with a probability in space and velocity. Quantum mechanics describes the temporal development of this probability as the consequence of inertia and forces.

One of the most important insights of quantum mechanics is the uncertainty of some physical quantities. These cannot be exactly measured at once at a given time. For example, the location of an electron, of which the velocity is known with an accuracy of 1 percent of the speed of light, can never be measured more precisely than the diameter of an atom. The reason for the uncertainty is the indefiniteness of the quantum mechanical reality discovered by Niels Bohr and Werner Heisenberg. Due to the uncertainty of the development, a quantum system is predictable only in form of a probability. In the past, the role of uncertainty was expected mostly in the realm of elementary particles. The importance of quantum phenomena in everyday life becomes increasingly obvious. They determine, for example, the chemical and physical properties of molecules, play a role in the transmission of genetic information by organisms, and allow technical applications in electronic components.

THOMAS: So the smallest elements of nature stay hidden behind a curtain. Behind this quantum veil, everything looks nebulous and blurred. I still don't understand how clouds can escape the depth of a well.

SHELDON: The cloud you're wondering about isn't real. It represents only the probability that one can find the electron at a specific location. Sometimes it behaves like a wave that sloshes over a wall. If the electron encounters such a wall, one part of the probability flies farther; the other is reflected. In our macroscopic world, we interpret this as pure chance whether the electron flies on or not.

THOMAS: What determines whether it acts now more like a wave or like a ball?

SHELDON: The observer decides. If the particle isn't observed, it behaves like a wave in three-dimensional space. In some observations, the probability wave breaks down to a point where the electron manifests itself as a particle. Niels Bohr talked about complementarity in view of the double nature of the electron both as wave and as particle. Mathematically and technically, both descriptions rule each other out. The nature of the electron is determined only when it's observed. The observation is an irreversible interference in the world and only then the world becomes real.

THOMAS: How strange! If I don't look at it, the electron plays the role of a cloud, but as soon as I look at it, it suddenly changes into a small solid ball or a wave. Now how about all those processes that aren't under human observation? Nature hardly waited for the experimental physicist to unveil the components of complementarity. Even without people, there must be processes in which the probability cloud freezes to a solid particle.

SHELDON: You've uncovered an unresolved question in quantum theory. No matter how well its mathematical formalism is known and confirmed, its interpretation remains just as vague. We scientists behave like computer users, who can use the computer as a tool without knowing how it really works. The role of the observer is unclear and contested.

THOMAS: One of my theology professors once advocated the bold assertion that just as man is the observer in the microcosmos, God himself is the great observer in the universe. He lets the multitude of possible worlds crystallize by his sight to one real world, much as a physicist permits by his observation the probability distribution in form of a cloud collapse into a particle.[1] With his creative vision, God not only created the universe out of the foggy realm of possibilities but also works and is omnipresent in the world.

SHELDON: A daring notion! But as long we haven't understood what's actually going on with the observing person, we shouldn't already draw a conclusion about God's creative activity. In physics we study the smallest components of matter. In doing so, it's unavoidable that in the subatomic world the physical instruments influence what they're monitoring. For purely technical reasons, there's no observation without active manipulation. The observing instrument works also with particles or waves, which are subject to the uncertainty principle of quantum mechanics.

THOMAS: So you admit that the most objective and exact science claims that perception contributes creatively to the construction of the world.

SHELDON: That's right. And that's also the case in our central nervous system, where, for instance, specific areas of the brain convert rather diffuse signals from the retina to clear optical perceptions.

At this point Hoihong Wong interrupted.

"I am not a physicist, so I asked Professor Harald Rightman for his independent opinion of the issues the astronauts have been discussing."

On the Interpretation of Quantum Mechanics

My dear colleague Sheldon Cutter appears to interpret quantum mechanics according to the classic Copenhagen interpretation, as many physicists still do today out of sheer convenience. They postulate that reality forms in the moment when a person consciously notes the result of their observation. Niels Bohr in Copenhagen, Denmark, and the German Werner Heisenberg proposed this version in order to make the quantum mechanical indeterminacy self-consistent.

This interpretation is so anthropocentric, however, that it would not be generally accepted today. It may be useful for explaining laboratory experiments, but when it's applied to the universe, it leads to the absurd conclusion that everything that is not observed does not exist. Under those circumstances, humans and human consciousness could never have developed. The observer cannot be the producer of reality if the observer him- or herself is not already real.

The interpretation of quantum mechanics is by far not as important as the New Agers, including Thomas Haubensak, still believe.[2] The mathematics can be used without any interpretation and produces exact results, as in the case of quantum well diodes. According to today's view, there is no need for an observer to bring the probability distribution to collapse. It's sufficient when, for example, the location of the particle is manifested irreversibly and macroscopically by any process. This happens with the detector's counting a photon, with the blackening of a film grain, or by the mutation of a gene.

A conscious observer is not needed. They can detect these irreversible manifestations afterward, however, and take a reading. Each appearance of reality is an irreversible act that lights up the actual things like a stroboscope. Every new picture captures reality and unveils the temporal course of events. It is not the human consciousness that generates reality but rather the sum of these continual irreversible processes in time.[3]

Quanta Knock at the Door
A Postcard from Saturn for the Cyber School

Hello, Earth!
In my free time I sometimes close my eyes and try to imagine my place in the solar system. More than a light-hour separates us from Earth. Saturn and Titan, huge in our portholes, become small within the boundless and empty space around us.

I can see, even with my eyes closed, that space is not totally empty. When I close my eyes, flashes of light flare up and silently explode in bright colors. The old moon astronauts in the 1960s knew this phenomenon and explained it as being caused by elementary particles, mainly protons, accelerated by the sun or from the depth of the universe. The tiny particles speed along, penetrate the cabin walls, and trigger light effects on the human retina.

I think that the defect in our transmitter was caused by the increase of these occurrences. After all, we're in the midst of the powerful magnetosphere of Saturn, which reaches far beyond Titan's orbit and creates a sheet in which an electric current circles Saturn in the extension of the rings' plane.

Hermes Trismegistus floats in this current. Magnetic fields encompass our spaceship and electric fields speed up particles to enormous energy in close proximity to us.

The spectacular light show in our eyes is disconcerting. We perceive a picture that does not correspond to reality. The blinding flashes happen only in our retina. When I then shout, "Did you see that?" Thomas looks astonished and worried at me.

That we "see" single elementary particles shows how close the quantum world is. Space outside is not empty; hosts of particles take their course accelerated by surges of enormous electrical and magnetic forces.

From Saturn with best regards, Sheldon Cutter

And Endlessly Surges the Quantum Ocean

THOMAS: Enough said, Sheldon. It's time for our daily fitness program.

SHELDON: Right. This weightless life and lack of physical activity isn't good for us.

THOMAS: These adhesive shoes are great. Do you remember how we learned early in our journey to loop around the living quarters? I push off at an angle to the round wall and land a little farther on it again. The centrifugal force always brings me back to the round wall.

SHELDON: Since I'm a former pole-vaulter, I can do even ten rounds a minute. Yes, another victory of science over religion.

THOMAS: Ha ha. Nevertheless, the looping run produces a centrifugal force that exceeds the gravitational pull of Earth and gives me, too, the feeling of standing once again on solid ground.

SHELDON: I'm going to work on upper-body strength with elastic ropes in the fitness area. Be back in five.

THOMAS: That was more like two. You must be getting tired. But never mind, I just saw another one of those fata morganas. I used to think outer space was totally empty. Now it seems it's full of particles and quanta.

SHELDON: There's no empty space in the physical world, actually. Even this empty space we call "vacuum" is hubbub. Elementary particles and field quanta continuously form and vanish.

THOMAS: How can elementary particles form and disappear? They are the most stabile components of matter.

SHELDON: They're not fundamental. Energy is still more basic than the particles.

THOMAS: But can there be energy in empty space?

SHELDON: By definition, the vacuum represents the lowest energy state.[1] According to quantum theory, however, energy is uncertain and thus blurred, and has to fluctuate constantly around the minimum value, the so-called zero-point energy. These undulating quantum fluctuations are local transitory fields of energy. At places with higher energy, they act like particles that form for a short time and disappear again.

THOMAS: Then the ancient philosophers were right. There is no totally empty space! How am I supposed to imagine such a pseudo-nothing?

SHELDON: It's like the chaotic fluctuations of the rough, choppy waves on the open sea. These vacuum fluctuations are incredibly fast in our everyday macroscopic circumstances. Even for the energy of the tiny mass of an electron, they only last some 10^{-21} seconds. Their size is less than a billionth of an inch.

THOMAS: The image of the ocean with its waves is impressive! Do energy and matter develop from this oceanic basis?

SHELDON: Yes, in a way. These particles, however, must disappear again within the temporal uncertainty in order to comply with the law of conservation of energy. Their life span is so short that they can't be observed, which is why we call them "virtual" particles. They're similar to dolphins playing in the sea. As soon as you have your camera ready to snap the photo, they've already disappeared. A cosmic "censor" ensures that one can never take a picture of a virtual particle. The larger the energy of the particle, the shorter its life span. It borrows its energy needed for creation from the vacuum, but it must pay it back before its absence is noticed.

THOMAS: I can imagine a stringent high archangel in charge of all the divine books containing the laws of nature. He monitors their observance. When he closes his eyes for a short time, the world of virtual particles opens up a wondrous dance of the elves. When he looks again, everything is once more adjusted into the orderly world. A truly remarkable phenomenon at the foundation of reality! Does it have an effect in our world?

SHELDON: And how! It was first detected directly in spectral lines of atoms. The course of electrons around the atomic nucleus is slightly disturbed by the presence of virtual particles.[2] Their shadowy dance has real consequences. It's measurable and therefore a comprehensible offshoot of the great quantum ocean. Nevertheless, it's a tiny effect that can only be proven by sophisticated devices. The quantum vacuum probably concerns a reality that includes far more. The energy that is everywhere, and even in the vacuum, causes a pressure in the entire universe and bloats the expanding space. Most physicists believe that this zero-point energy is the dark energy that acts like anti-gravitation and accelerates the universe to ever faster expansion.

THOMAS: A stunning effect! And still we notice remarkably little about it.

SHELDON: Quantum fields show up most directly when energy is supplied externally. Virtual particles can then suddenly become real free particles that aren't tied up anymore with the temporal uncertainty of the vacuum. In this way, real particles develop out of virtual particles.

THOMAS: Stop for a moment. I want to do the abdominal and back muscles to strengthen my back.

SHELDON: Yes, I'll do those as well.

THOMAS: Enough ... I'm out of breath ... But something is still a mystery to me. Is the existence of the solid "real" caused by the swaying and ebbing away of this virtual ocean? Does it swim in the vacuum like an iceberg?

SHELDON: The image of the real world as an iceberg requires an addition, since the surface of this glacier isle itself fluctuates like a lake and is interspersed by virtual particles. The realm of the virtual permeates not only the vacuum but also the world of real particles, which are only put onto the vacuum. Matter isn't motionless, but mercurial energy. It exists in form of wave fields, either in the form of the lowest energy state, that is the vacuum, or as a higher excited state—namely, the particle.[3] Matter constantly tries out all types of states that are possible by the basic laws and symmetries, and never keeps too precisely the rules of energy conservation while hiding.

THOMAS: The realm of the virtual seems to penetrate the realm of reality so much that it continually recedes and newly comes out from it again.

SHELDON: The real and virtual worlds are indeed very close. Real and virtual electrons have also the same mass and charge. It needs only a pinch of energy and suddenly a particle from the infinite virtual stock becomes real. In reverse, real particles constantly take up a virtual state for a short time.

THOMAS: I wonder to what extent the view of the quantum physicist in some areas isn't also to be interpreted with the return of the ancient *mythical cosmologies*—cosmologies where what exists is based on a primal ground that brought it forth, which at the same time constantly threatens it. The ancient Egyptians believed that their world was totally embedded in the boundlessness of the pre-cosmic chaos. Every evening, the sun returned into the creative depth to rise renewed in the east in the morning. Being sleeping or drunken, the living creatures let themselves be imbued by the reinvigorating power of the primeval ocean that carried the whole creation. The symbol for this was the flooding of the Nile, annually returning the waters on the land to grant new fertility.

SHELDON: The real world disappears consistently into the virtual and instantly reemerges. It may indeed appear chaotic because it isn't predictable in detail. What you're saying about chaos from mythology, however, is different and seems to have mainly negative associations.

THOMAS: And rightly so! In the ancient cosmologies, the primal ground was ambivalent. Inexplicable chaos brought not only renewal and blessings but also death and destruction: tidal waves, diseases, political enemies threatening the people. Life and death were inseparably intertwined here with each other.

SHELDON: Your analogy falls short at this point. The chaos in the virtual world isn't either positive or negative. Values appear only with humans. I can understand how this omnipresent zero-point energy seems weird and threatening for you. But it's not at all the same as the mythological primal basis.

THOMAS: But something common remains. All that exists rests on a foundation filled with tremendous potential. It confirms my belief that an unknown divine power creates everything in the universe.

SHELDON: Here's a bit of expert commentary about quantum field theory. I'll project it on your monitor, Thomas. The text is from Professor Curt Schuhschnabel, a colleague of mine from the University of California at Berkeley. I just had it on my to-do list for reading.

The Energy of the Vacuum

I thoroughly agree with my physicist colleague Sheldon that there is energy in every volume of space—one of the most peculiar and incomprehensible statements of the quantum field theory. Even in a vacuum box, the minimal energy state of the space without real particles, and even if the temperature of the walls is at the absolute zero point, at -459.4°F, the vacuum still has a certain energy density, an almost constant value that does not need to be zero. The ocean of zero-point energy was confirmed by laboratory tests, but it still contains many mysteries. It's like we're in a boat rowing on the sea. The ocean can definitely be felt, but we know nothing about its depth.

It's surprising that the density of the zero-point energy is not just zero. It is embarrassing for me to admit that our equations result even in a humongous energy density, which is totally impossible. Either the infinities neutralize each other exactly or there is still a deficiency in the well-proven equations. It's also possible that the zero-point energy has something to do with the virtual particles or their radiation. Perhaps it stems from the time of the beginning of the universe. The quantum fluctuations of the primeval time have certainly long since vanished, but their virtual powers could still have left behind a trace in the form of fluctuating fields and energy.

Since the 1990s, there are increasing indications that the expansion of the universe is accelerating.[4] We generally assume that the pressure of the zero-point energy manifests itself on the space-time continuum in the whole universe. If so, its effect exceeds the opposing gravitation of the galaxies so that their expansion of the universe accelerates.

The two astronauts find themselves on extremely speculative paths, however, where they connect the vacuum directly to the universe and therefore make it the origin of all becoming. There is, of course, the well-known hypothesis of the origin of the universe out of the vacuum. The primeval vacuum before the formation of the universe would differ from today's vacua that there was no surrounding matter that could be used as reference for place and time. Without particles or photons, space and time cannot be defined.

Thus we cannot talk about symmetries and developments. There was no physics in today's sense. The first particles opened up space and time. Nevertheless, the formation of the universe would already follow our presently known laws: by a fluctuation that would maintain certain quantum numbers and conserved quantities (for instance, energy) within their uncertainty. The primeval vacuum may have already contained all physics of the later universe. The formation of matter would then take place according to these laws.

Even if I am not aware of a better speculation about the beginning of our universe, I consider this kind of physics to be pure dawdling. How can we hope to ever prove these statements in an experiment?

Randall commented here: "When the main transmitter of the *Hermes Trismegistus* fell silent a billion miles from home, we all worried about the survival of the astronauts. But out at Saturn, this breakdown opened a window into a new reality for Thomas Haubensak. Look at his "diary"!

Dance with the Floods
From Thomas's Diary (April 17)

What a fascinating gate opened for me today! Sheldon has explained to me that in his scientific world of physical reality, we don't find solid atoms chained into a rigid lattice, as I was taught in high school many years ago, but rather a swirling dance of particles forced by a guardian angel under the noble law of necessity and the conservation of energy.

This game of an exuberant creativity at the foundation of all being is breathtaking. It's taking place in a virtual kingdom that doesn't deserve to be called reality, but it is more than just an empire of ghosts and unreal possibilities. The figures of the real world rise, then, like icebergs out of the endlessness of the surging ocean of virtual matter and energy at the level of elementary particles.

Fragments of thoughts when awaking, associating, or decision searching remind me of the zero-point ocean in my own brain. Like ice floating in the sea, shreds of thoughts surface and instantly sink again. An inspiring idea flashes up, a whiff of euphoria, a shadowy recollection—and then everything is gone again in the cerebral surging wave. Even the clearest thoughts of the day: Don't they also tend to dissolve at short notice and form again? Are the waves of the primeval ocean breaking against the rocky cliffs of my daily reality?

Maritime images play an important role for the American philosopher and psychologist William James.[5] According to him, our souls rest like isles in a wide ocean, a continuum of cosmic awareness, with which they communicate in special states of consciousness. From this, James allowed himself to be led to a "panpsychistic" perception of the universe, claiming that all objects have a mind-like aspect. Our discussions today have driven me far in this direction. I suspect that the physicists' equations actually mean something much different than the mythological chaos or the depth of the subliminal consciousness. Nevertheless, the vision of a universe fascinates me, in which the structures and figures rise up out of an unending ocean of flooding energy that has not yet crystallized into well-defined reality in time and space.

The immense apparent emptiness of space that we're surrounded by raises many questions for me. Every being in the universe seems to be enveloped with a dense cloud of possibilities, from which it constantly refreshes itself. Do this ocean's breaking waves have an effect on our virtual but perhaps not actual world? Or does the sublime archangel whom I call God and who watches over the strict cosmic energy laws prevent the encroachment of the enchanting quantum dance?

Can we communicate with this surging ocean? Are there human experiences in which we'll be invited to dance in this cosmic roundelay? Are meditation, prayer, art, or love gates into this swirling basis of being?

Creation without Interruption

SHELDON: I just read, at your request, your diary entry from yesterday. You seem fascinated by the maritime imagery. I also love the sea and especially the Indian Ocean, but I can't share your enthusiasm. Why is the "quantum ocean" important to you as a theologian?

THOMAS: The attraction of oceanic images for me is connected with their meaning in the field of mythology. As a student, I was already fascinated by the ancient Babylonian epos regarding the origin of the world, where the formation of the world takes place by killing the chaotic primeval snake Tiamat. The god Marduk forms heaven and earth out of its body, which had been divided into two pieces. Homer names the Titan god Okeanos the origin of all. From him descend all seas, rivers, springs, and the cool winds. The first "modern" philosopher, Thales, takes up this idea when he identifies water as primary matter.

SHELDON: The speculative world of Thales and his theory is based far more on mythical images than on objective observations in nature. Modern cosmology vehemently dissociates itself from these kinds of speculations. But I don't want to unsettle you with a précis on today's theories about the formation of the universe.

THOMAS: No, no . . . I hope we'll eventually debate cosmology in detail. Antique texts on creation aren't only concerned with the beginnings, but above all with the continued existence and the nature of the present world. They teach about the mythic ocean with its creative and threatening power in daily life, in the secret of fertility, and in the horror of flooding. Myths of the origin are aimed at the present time. They interpret the order of the world and offer guidance on survival to us humans.

SHELDON: How can you be so romantic in our third millennium? Modern cosmology answers the old questions and has distanced itself from its prescientific myths. Today's empiric-mathematical approach fundamentally differs from the former intuitions. A chasm yawns between modern quantum theory and the ancient Egyptians' and Babylonians' waters of chaos.

THOMAS: I know, Sheldon. Believe me, every time I look out the porthole, the distance between us and those times dawns on me anew. The ancient oriental world model, formed like a bell-shaped glass cover, dissolved into the infiniteness of the universe. On the other hand, there is also the return of old insights. The classic Greek negation of empty space, *the horror vacui*, is confirmed today again. That's why I believe older views regarding the divine conservation of the world also deserve more attention today.

SHELDON: Preservation of the world? Do you think that an external transcendent power is needed to conserve the universe or matter?

THOMAS: Many ancient mythological cosmologies revolve around the preservation of the world from chaos and collapse, constantly accomplished by divine powers. They come and go in antiquity, and have changed their identities and roles.

In the Bible's priestly creation narrative, the world's existence, however, is ensured by the continuous presence of God. God once and for all laid the world's foundations, and there's no longer any need for ceaseless repression of chaotic powers. In platonic philosophy, the cosmos takes part in the divine eternity and permanence. Science of the early modern age finally gets along without God.

SHELDON: Indeed! Laplace is said to have answered Napoleon's question about which role God plays in his world system, with the assertion that the world maintains itself.

THOMAS: Even I've heard of Pierre-Simon Laplace.

SHELDON: Yes, the "French Newton." He was a famous eighteenth-century scholar and mathematician who'd proposed his nebular hypothesis of the origins of the solar system. Like the inertia of our spaceship continues its present motion, the process of the cosmic development progresses by itself.

THOMAS: Newton, the founder of modern physics, located God's activity in gravitation. Today's physics ties gravitation to the curvature of space. Newton, however, proclaimed the necessity of a continuing conservation and balancing of gravity so the universe doesn't crumble into chaotic disorder and immobility. That seems to me like God working continuously in space and time. The clockwork of creation must be repeatedly serviced.

SHELDON: Science has gone far beyond Newton on this point. The nineteenth century at the latest recognized that the world clock functions excellently and has no need for further intervention by the cosmic watchmaker. The twentieth century scrapped the mechanical model of a clock, but learned that complex structures like cyclones, stars, chemical patterns, living creatures, and societies can form themselves without an external law enforcement agency.

THOMAS: My interest in the discovery of the quantum ocean is something different. It seems that our macroscopic world lives from an almost never-ending source, which underlies all real being and keeps the universe from collapsing.

SHELDON: And you want to equate this background with God?

THOMAS: No, I'm once again just hunting for meaningful analogies. A good old stream of theological tradition tries to read the world as a parable for God.[1] Is there a cosmological variety of this reading? In the same way, as the universe rests on an ocean of virtual energy and is permeated by it, creation is carried by God and conserved by him.

SHELDON: The equation doesn't work. The vacuum contains the universe but doesn't maintain it. The universe doesn't need a subsistence from the outside. The first law of thermodynamics states that energy can't vanish.

THOMAS: Imagine that an extra-cosmic daemon drinks up the whole vacuum. Could the universe still exist on its own?

SHELDON: No, it must disappear at the same time along with the vacuum.

THOMAS: Well, here's an interesting analogy. In theological cosmology, God preserves creation by his constant creation of space and time. Without his incessant work, the world would immediately disintegrate. In physical cosmology, the universe is embedded in the surging vacuum as its breeding ground. So isn't this basically the same? The surging ocean at the foundation of the universe reflects the exuberant creativity of the divine cause from where creation perpetually receives its existence.

SHELDON: Be careful! You're fishing for analogies in the quantum ocean. The quanta's seemingly wild dance is subject to extremely strict laws. In the macroscopic world of particles, atoms, molecules, and living creatures, this fluctuation has almost no effect. There are real particles in the handles of our spaceship, on which you bump your head while floating gravity-free. The virtual particles, on the other hand, have a shadowy existence in our reality.

THOMAS: In the religious perspective, God preserves his creation by continued new work. When in physics, even the hard matter is subject to incessant fluctuation without losing its identity, how much more does it apply for creation continuously conserved by God.

SHELDON: That's a big one! According to your view, on one hand, God no longer intervenes in the world, so science cannot get a hold on him. Thus you voluntarily lock the back door that some have held open for him till now. On the other hand, you lead him through the main entrance, make him the indispensable keeper of the world, and invoke even the latest physics.

THOMAS: It would be poor theology if it would assign God only those locations where science doesn't yet have an adequate explanation. God would become a stopgap and his latitude would become narrower with each advance in science. In the religious perspective, it's the reverse: God fills everything; his splendor manifests itself even where science can explain everything with natural laws.

SHELDON: So why do you take such an interest in physics?

THOMAS: Because it's about the perception of the world as a creation. In this inner perspective on the cosmos comes into view what the sciences try to describe from their outer perspective.

SHELDON: Oh no! We've arrived again at your "third eye." What has that to do with parables? Does the third eye see by the means of analogies?

THOMAS: You play another metaphor into my hands. Even the third eye doesn't get to see God's activity directly. Creation is full of mirrors, in which the divine light flashes. The world may have been one perfect mirror at the beginning, which then burst into thousands of fragments. Now they all reflect the one divine light in their own way. The analogies I'm fishing for are those kinds of mirror pieces.

SHELDON: Here, too, you could fall back again on modern cosmology. In the first fractions of a second of the big bang, perfect symmetries ruled between the elementary natural forces that shortly afterward shattered. However, we discussed before that it does not really make sense to use such abstract theoretical constructions like quantum field theory or the laws of symmetry for metaphors. As far as I know, Jesus spoke about simple things.

THOMAS: Sure! His world was the one of farmers and fishermen in Palestine. The beauty of the wild lilies, transitory and quick to wither away, reflected for him something of the exuberant magnificence of God.[2] Why shouldn't it be possible to use insights from high-tech particle accelerators and telescopes stationed in space as metaphors for creation? The pictures from JWST,[3] the *Hubble* telescope's successor, inspired me in my youth. I've never understood the reservation toward scientific discoveries displayed by some of my theological teachers. They only trusted the immediate everyday appearance, but not distanced science. I'm not so sure about this segregation of reality.

SHELDON: On the other hand, you are fascinated by vivid and easily understandable images of our findings. The problem is that there is a long and convoluted path that leads from technically demanding experimental setups and complicated mathematical delineations to illustrative descriptions of the world. The image of the cosmic quantum dance only approximates what can hardly be conveyed in words and images.

THOMAS: This problem also comes up with theology. If we want to talk about God, we are confronted with the unsurpassable limits of the conceptual language. Yet theology won't simply let God sink into the unspeakable and the unthinkable. With the help of metaphors, parables, or analogies, it attempts to talk about the inexpressible. It's important to me that these forms of speech are more than rough crutches and poor substitutes.

Our debate reminds me of a classic scientific dispute between Newton and Leibniz. It was fought with weapons of metaphors that we can still appreciate today. Here's a short summery you may read tonight.

God of the Workday and God of the Sabbath

During the early eighteenth century, a battle began between two giants of intellectual life, Sir Isaac Newton, the founder of classical physics, and Baron Gottfried Wilhelm Leibniz, the ingenious polymath and philosopher. Both are regarded as the inventor of infinitesimal calculus. They conducted an embittered quarrel about its primary discoverer.

More importantly, their theological approaches clashed with one another: Newton's God acts in the world like the biblical Creator in the first six days, but he also maintains and preserves it as a whole. The God of Leibniz finished his work by creating the best of all possible worlds, and rests on the seventh day. In a letter addressed to the Princess of Wales, Leibniz expresses his disapproval about the teaching of Mr. Newton and his followers, whereupon from time to time God must clean and wind up the world watch anew, since he did not understand how to make the clockwork perfectly from the beginning.

Newton consequently encouraged his loyal supporter Samuel Clarke, court chaplain in London, to a counterattack. It came to an extended increasingly fierce correspondence, which strides from the understanding of freedom to concerning the problem of miracles, as well as the relationship of God to time and space.

Clarke, Newton's spokesman, claims that Leibniz's conception of a perfect orderly world says goodbye to God and opens a gate for atheism. The dispute concentrates especially on the importance of gravitation. For Newton, the cause of the gravity force is not in matter itself but to be sought in a nonphysical spiritual power. He indicates in a disguised way what his supporters then loudly proclaim: Gravitation is a proof for the existence of God. One day in the distant future, God would have to prevent the disintegration of the universe by its own gravity.

In contrast to this, Leibniz establishes that Newton's God apparently lacked so much in foresight that he "lives from one day to the next," although he has in reality made provisions in advance.

The exciting part in this debate is a peculiar reversal of the fronts that subsequently appeared. At the end, Newton was acclaimed to be right with his mathematical understanding of the physically described world, while Leibniz still worked with ancient and scholastic ways of thinking. As soon as gravitation became a purely physical quantity, the working God of Newton had to withdraw into the supernatural, inactive position, which had been reserved for him by Leibniz. Henceforth, the world clock needed no further divine service.

Why So Uncertain?

Encouraged by the methodological discussion, Sheldon presented some thoughts of his own concerning a modern topic.

The classical physicists at the beginning of the twentieth century, first and foremost Albert Einstein, felt the uncertainty of quantum theory a grievous flaw. The indeterminacy hinders exact forecasts and destroys the image of a cosmic clockwork.

Why is the world the way it is? Does the uncertainty perhaps have a purpose? What would the universe be without the indeterminacy of the quantum world? Is there a cosmic plan behind this? Physicists are particularly interested in the how. *Let us ask once* why *only a measurement brings forth reality!*

The quantum world includes also the equally crazy feature of nonlocality. When, for instance, a pion—a subatomic particle—decays into an electron and its antiparticle, a positron, both particles fly in opposite directions and both rotate around themselves. Their spin must sum up to zero because a pion always has a zero spin, which is conserved and must add up to zero for the two daughters.

If the electron's spin regarding a certain direction is clockwise, the positron's spin must be anticlockwise, wherever it is in the universe. Both particles are "entangled" and their probabilities are intermeshed. If I know the property of one, then I know the other. Electron and positron remain connected with each other in a mysterious way and form a nonlocal *unity.*

Like other quantum numbers, the spin direction is indeterminate until measured. When I'm observing here near Saturn one particle's spin, its probability distribution collapses at a certain value. At the same moment, the spin of the twin is also determined, which, for instance, could be on Earth.

Whether one measures it just then or not, does not matter. Its spin is freed instantly from quantum uncertainty. From this moment on, the spin is "sharp," and the earthly particle reacts correspondingly. My measurement here, thus, has an effect on Earth, which went there immediately and faster than light.

What about telecommunication at superluminal velocity? How practical it would be if we could use this effect to chat with those on Earth without always having to wait two and a half hours for an answer. The hopes for using quantum nonlocality for communication unfortunately were dashed. It would have been too nice.

Or perhaps not! Forces play the main role within the cosmic play of cause and effect. They are transmitted by field quanta, such as photons, gluons, W or Z bosons, which at the most move at the speed of light. Even gravity adheres to this rule.

The causality of the whole cosmos is based on the limited signal speed. With instant interactions, cause and effect would be simultaneous and could not be differentiated. It's the core of Einstein's theory of relativity that causality is not possible when two events are simultaneous for the observer. In a spaceship traveling at superluminal speed, even the sequence of cause and effect could reverse. The wonderful play of cosmic development would be greatly confused with instantaneous transmission of information and would not be thinkable at all.

It is exactly the indetermination of quantum mechanics that hinders the misuse of nonlocal unity for communication. For example, I may want to use the electron's spin here in Saturn's system in order to transmit a message via an entangled positron kept on Earth. As soon as I manipulate the electron's spin and bring it into a certain direction to upload some information, the electron and its twin lose the quantum uncertainty. They are no longer entangled with each other and become useless for communication. The nonlocal unity of the pair of twins exists only under the cover of uncertainty.

The quantum mechanical uncertainty must be a central part of the cosmic design. Relativity theory dissolved the universe into single particles. These particles are isolated from each other. Quantum entanglement *connects these individual points again into a whole. Einstein rejected both quantum uncertainty and entanglement, but together they are compatible and agree with observations.*

Nonlocality can also occur between more than two particles. The typical quantum condition of a group of many particles has nonlocal properties. One may ask which parts of the universe are entangled and form a unit. The reduction to single particles is very successful in physics so far and cannot be totally wrong. Thus I cannot believe that the whole universe is entangled.

Physics or Mysticism?
From Sheldon's Notebook (April 18)

Recently Thomas asked about the fundamentals of quantum mechanics in mystical holistic experiences. I sidestepped the subject, since New Agers and telekineticists have fantasized about it for decades. But who knows exactly what takes place in the human brain with its high-molecular processes? The long way from the entanglement of two elementary particles to the neurophysiological processes is still almost completely ahead of science.

It seems to me wonderful enough how subtle the uncertainty of the quantum world reconciles the hard contrasts of limited speed of light and nonlocation, hindering a contradiction. That's why there is uncertainty! The coexistence of relativity and quantum entanglement does not compulsively follow from the basic equations, but this is the only way both fundamental theories in physics make sense.

THOMAS: Thank you for those two edifying writings.

SHELDON: OK. Enough said. We've now gone far from the equations of quantum physics! I became really uncomfortable in those days when New Age enthusiasts boldly mixed Far Eastern religion and selected elements of particle physics to a Tao of Physics. Some succumbed to a very incidental consonance of two concepts that originate from totally different contexts.[1]

THOMAS: I was also suspicious at first of the disciples of the Quantum-Tao. But let me ask the question: Is it conceivable that the wise men of the Far East by virtue of their minds, trained through decades of meditation, reached a direct insight in the deep dimension of reality that today's modern science will discover by a totally different route? If one proceeds from a unity of reality, such convergences may occur.

SHELDON: If the Tibetan hermit, the dancing dervish, and the experimenting physicist celebrated one and the same truth, a beautiful picture would result. If this were true, we could save a lot of money and effort. The enormously expensive particle accelerators of the latest generation would have been totally unnecessary. Instead of the gigantic Large Hadron Collider in Geneva, a thousand meditation centers could have been built. But as a physicist, I'm certain that a yogi can't discover the vacuum with its virtual particles, any more than a botanist can ever find a planet.

THOMAS: Why shouldn't the human mind be able to establish direct contact with the foundation of reality? I'm thinking of an intuitive awareness of existence that doesn't segregate reality into separate objects from each other. Doesn't today's physics conclude that nature can't ultimately be separated into isolated objects? The inward contemplation practice by Buddhists aims for a nondiscursive communication with the absolute being that leads beyond the duality of subject and object.

SHELDON: Watch out for conceptual obfuscation in the interpretation of quantum mechanical issues. It's OK to say that during an experiment the observed object and the observer form a coherent system. Our target is not nature per se, but nature that is subjected to a certain question. The observations result from a certain measuring setup with ingenious instruments, and they are interpreted through an extremely precise formalism. But that has hardly anything to do with poetical statements of religious texts and nothing at all with diffuse allegations of holistic enthusiasts. By the way, I was never enthralled by Eastern spirituality that spills in regular waves over into the West. I can understand even less why you—as a traditionally-minded Westerner—set out into this mumbo-jumbo garden.

THOMAS: Let's stay with cosmology. I argued that God ensures the continued existence of the world and its laws with a continuous creating out of nothing. Everything continuously sinks back into nothing and is again called into being. A whiff of this continuous creation touches me in my meditation, where I participate in the large disintegration and renewal with every breath. Ostensibly, the world seems to be stable and firmly established, but it dissolves into a cosmic universe in the practice of concentrative "sitting."

SHELDON: Doesn't that boil down to explaining the world as an illusion? Quantum physics restricts the instability of the particles to extremely short intervals. The larger the fluctuation energy, the shorter the interval. The actual reality of macroscopic objects that is our everyday world won't be affected at all.

THOMAS: I'm not interested in denying the relative stability of our world. I'm too much of a Westerner to experience it just as phantasmagoria or veil. Yet I'd like to shake the dullness with which we meet reality. We give it quite the status of something unquestionably existing. We lock it up in the prison of isolated objects and compact substances and treat it accordingly. A veil actually sinks down between us and reality.

SHELDON: Then I can only hope that you won't take the boldness to misuse the veil that cloaks the quantum world. An enthusiastic columnist once equated it with the ancient Indian maya, the illusory deception of the phenomenal world.

THOMAS: Not even the enraptured sphere of Saturn could make me a quantum fool who snatches rashly at everything that looks similar from a distance.

SHELDON: I'm glad we both profess an enlightened rationality, but you refer to personal experiences for which I have trouble emphasizing. How do you get from meditative conditions to cosmological hypotheses? Do you try to extrapolate subjective experiences into worldviews that claim objective validity?

THOMAS: No, this time my concern is more modest. I'm thinking of two drilling teams, working from opposite sides of a mountain to meet in the middle. There's the spiritual exercise leading toward the inside: In contemplation I feel as if I'm held over an abyss. With dedication and awareness, I take part in a constant movement of going away and coming back anew. In that movement I experience reality. On the other hand, there's the path of experimental physics directed to the external world. With astonishment and delight I realize that one can reach similar perceptions both ways. Whether the drilling teams actually meet or whether they dig past each other, I don't know.

SHELDON: They not only miss each other by a hand's breadth, but they may be drilling in different mountains. From a scientific view, there aren't even the slightest signs of instability of the universe that must be resolved by external factors. The universe is amazingly stabile with regard to the fundamental cosmic constants. The strengths of the elementary forces seem to have hardly changed since the beginning of time. The same also applies for the basic constants of the whole cosmos. The inertial motion of bodies requires no external cosmic support but obeys alone the laws of classical mechanics. Our personal amazement at the world and our own lives is a totally different thing, of course. Amazement surely has its validity here. The question is only whether one may draw far-reaching speculative conclusions from this emotion.

THOMAS: Cosmic and mental worlds must be somehow interlocked. After all, mental training is a very effective way to success in sports and it has to do with quantum processes in the brain.

Here's another entry from my diary that reveals my feelings about your quantum universe, Sheldon.

Our Mother, Thou Who Art in the Depth
From Thomas's Diary (April 19)

Sometimes I am jealous of our colleagues who are circling Saturn's moon Titan and will soon start the landing procedures. To set one's foot on another celestial body must be a tremendous experience. I like to imagine the excursions the group will undertake with the caterpillar. Some striking lookout points are on the agenda.

What an exotic adventure: mountain climbing on Saturn's largest moon! I can hardly wait for their trip to the coast, where the thundering waves of the methane ocean will meet them—an ocean with strange methane icebergs under a dark-red sky where the sun is seen only as a weak, diffuse blotch of light.

The unknown lures with sounds of sirens! The pictures from ORPHEUS *have only partially lifted the orange veil of haze spread out over the mysterious moon. A hidden world dreams deep under us. Does a seed of an archaic life still dwell down there?*

My powers of imagination glide from Titan's thick haze into the undissolvable veil that covers the entire quantum world. Our last talks about the quantum ocean and the peculiar laws that rule in this basal world still possess me. According to ancient tradition, Earth's ocean originated from Saturn's tears.[2] An old Gnostic school identified Cronus-Saturn with the force of the darkness of the abyss and the surging of the primeval waters—a power that "upholds the entities that abide, restrains what trembles, unlooses what is to come."[3]

I take a deep breath. How good that the biozone of Hermes Trismegistus *can produce fresh oxygen-rich air! Memories of numerous retreats come to mind where, together with other searchers of truth, I was introduced by an old Japanese Zen master to the technique of meditation, the sitting. We practiced redirecting our attention away from the restless thoughts whirling around us to that most basic movement of life ceaselessly taking place in our body: breathing. The entire fullness of life is in it.*

Above all, the master drew our attention to the mystery of exhaling, the deliberate act of letting go, down into the depth of the body. By releasing the breath, we would enter into the origin. To breath out—to vanish, entrust ourselves to the depth, liquefaction of crystallized forms.

How peculiar—the efforts of our expedition, the exhausting visit to Titan and its ocean, all this searching in the distance sidetracks the view away from the very closest, from the movement with which I constantly communicate with the Great Depth: breathing! Exhaling, I give myself into nothing; inhaling, I receive myself again out of the nothing. While "sitting," I become involved in a movement of dying and resurrection.

I noticed repeatedly that modern quantum physical descriptions can be compared to the ways of thinking of the Far East. When knighted, Niels Bohr selected the Chinese symbol for the complementary primeval forces of yin and yang for his coat of arms. The uncertainty principle reminds me of the Buddhist principle of the inseparability between object and subject. Above all, the physical vacuum reminds one of the basic Buddhist conceptions of emptiness, according to which the manifold colorful manifestations of the world are based on a "void of abundance." Is there a hidden connection with the knowledge of physics whereupon matter and energy have to be regarded as excited states of the ever-present vacuum?

Two days ago, I talked with Sheldon about the importance of analogies between religious experiences and natural scientific facts. The analogies have to do with the reality, because they were not constructed by us but, so to speak, they picture the blueprint of the world. The whole reality is based on a common grammar.

Sheldon, on the other hand, takes the "nominalistic" position: the analogies count only as more or less witty gimmicks. They relate to reality like arbitrary names for objects. They'd be nothing but mere ciphers. Sheldon might quote Johann Wolfgang von Goethe to say, it is "only a parable."[4] *But I would counter with what one of my theological teachers said: It is* even *a parable! Did the deity even donate the cosmos images and parables representing its own being? Are parables due to the creatively speaking God?*

SHELDON: That inspires me to present you my latest postcard to Earth.

Birth of a Titan
A Postcard from Saturn for the Cyber School

Hello, Earth!
Our mission to Saturn conveys unexpected and fascinating insights to us into the primeval history of our solar system. The ODYSSEUS *probe has already taken quite a number of photos of Saturn's eighty-two moons with a resolution of only a few yards. The colors and formations give us a good impression of the surface history, which also allows us to make good guesses about the activity on the inside.*[5]

From the trajectory deviation of the probe we can calculate the interior structure of the moons. Impact craters testify to external influences. The picture that emerges from the synthesis of all details shows a primeval world of ice from the early solar system, although the four billion years since then have left behind many traces.

Let's briefly review the birth of the solar system. When the high-density core of a dark interstellar cloud core collapsed 4.56 billion years ago, the conservation of the angular momentum forced the masses of gas and dust into the form of a rotating disk. The more the gravitational force pulled the colossal whirl together, the faster it rotated and the hotter it became in the center, where the sun would soon develop.

At this time, there were also small enhancements in the disk, caused through the coagulation of rocky boulders made out of dust. The enhancements attracted gas from the disk, but they lost the race against the faster-growing primeval sun. The world of Saturn formed from such a whirling enhancement in the disk.

Temperatures near Saturn stayed low enough so the ice mantle on the dust did not evaporate in the light of the primeval sun and would not be carried away by the solar wind. The surface temperature of Saturn's satellites never reached the boiling point of water, thus sparing their ice layers the fate of evaporation—in contrast to Earth and the inner planets, where, at the end, almost only the solid core of silicate and iron remained.

Saturn's world stayed so cold that even the frozen methane and ammonia molecules did not sublimate into space but remained captivated in ice on the moons and are still at our disposal today.

Saturn's little vortex had about five million years to form the planet and the moons. Due to Saturn's strong gravity, no moons could develop in the inner region. Still today, debris from collisions and volcano dust from the entire Saturn system gather into the well-known ring. With the exception of Phoebe, all major moons lie pretty well in the plane of Saturn's ring. That is where they probably originated together.

Only Phoebe and many of the tiny moons orbit outside of this plane and in the opposite direction. Phoebe does not belong to the family and presumably was caught later. All the moons differ considerably from one another in shape and composition; each one seems to have its own history.

Titan is bigger than all its moon siblings in the solar system. It formed out of the matter circling near its orbit.[6] *The temperature rose through the contraction so that water, methane, and ammonia liquefied. The heavier dust particles of silicate, iron, and other elements sank to the center and today form a solid core. Perhaps it also contains radioactive elements that warm up the core, like on Earth. The largest part of Titan's mass, however, consists of water. In the innermost part, the water is possibly liquid; a thick layer of hundreds of miles of ice towers over it. There the bottled up, easily melted substances such as methane or ammonia may possibly lead to volcanic eruptions, when they are spewed out like water in earthly geysers.*

Ammonia gas initially formed a dense atmosphere—with an unbearable smell! Fortunately for our Titanauts, over billions of years the sun's ultraviolet light almost totally cracked up the ammonia molecules. The hydrogen vaporized into space, the heavier nitrogen stayed and is today the main component of the atmosphere, just as on Earth. Water ice forms the "bedrock" of Titan's surface and is probably the substance of the mountains covered by methane snow. Methane hydrate froze out on top of the water ice, and liquid methane and ethane accumulated in the low-lying areas to form oceans and lakes. We cannot forecast the scenery of the surface of Titan. Its landscape must be incredibly different from what we know on Earth.

Comets frequently hit the moon in the early history of Titan. The surfaces of the other moons that are not shrouded by an atmosphere speak a clear language. In the first seven hundred million years, there must have been ten times more impacts than during the entire time since then. Every few seconds, ice grains and debris of comets hailed down onto the young moons. About once every thousand years, a big comet, several miles wide, crashed down, sometimes almost shattering the moons. Rain and storms have eroded Titan's craters. Mountains of ice cliffs remain as relicts of recent impacts. Their steeply rising, fissured contours tower over methane seas and acetylene swamps.

Interstellar dark clouds contain about the same amount of carbon monoxide as nitrogen. However, carbon monoxide is extremely underrepresented in Titan's atmosphere. We can only imagine that it combined with liquid water to organic substances. This indirect indication of liquid water on the surface may be related to the comet impacts just mentioned. They brought water with them and melted the ground ice. Titan's precious water fascinates us the most because it bears many possibilities for prebiotic chemistry.

From Saturn with best regards, Sheldon Cutter

In the Beginning Was the Vacuum

SHELDON: Speaking of Titan, the touchdown of the Titan expedition was originally planned for a date shortly after April 9, 2051. But the landing craft ORPHEUS is still orbiting around Titan, and its radar system is still scanning the surface for a suitable landing place.

THOMAS: Right. I wonder how Takeo, Sergei, Shadia, Pablo, and Nicole are feeling. Tell me what's going on.

SHELDON: Mapping of the moon's surface should have taken only eight days, half of Titan's rotation period. But contradictions in the measurements occurred that at first were not explainable. IASA did not want to risk an uncertain, unsafe landing as the consequence of inaccurate maps, so they've postponed the landing day.

THOMAS: Well, this delay in the schedule gives us time for a new topic. I've been wanting to talk to you about the origin of the universe for a long time. In the ancient Near Eastern conceptions of the world, the cosmos was formed out of chaos. This is also reflected in the biblical creation narrative, where the dark primeval flood filled everything before God called light into existence. Would it be possible that the whole universe originated out of the quantum ocean? Could a real universe form out of the sea of virtual particles?

SHELDON: OK. Let's imagine that the universe originated from two virtual particles in a quantum fluctuation that emerged from the original vacuum. They moved away from each other as a result of a tunneling process, and finally became real.[1] In the quantum mechanical tunnel effect, all conserved quantities ultimately stay constant. Even so, the result was a new quantum state with incredible energy density. This subatomic mini-universe formed the core of the universe's matter in less than 10^{-31} seconds, perhaps passed through an inflation phase, and has been expanding ever since.

THOMAS: But earlier you insisted that energy can't increase. Where did this immense energy come from so suddenly?

SHELDON: Gravity's energy is negative. As soon as both particles sensed their mutual gravitational attraction, negative energy existed. It may have equaled the masses' positive energies and their energy of expansion. In fact, we observe today that the sum of all energies in the universe is about zero. This seems to be a strong indication that the universe originated out of nothing. It didn't need any input. The universe is a spontaneous fluctuation of the primitive vacuum. Colleagues of mine called it an "ultimate free lunch," because no price for energy had to be paid.[2]

THOMAS: Just what theology was waiting for! A basic tenet of theology is that creation is an appearance of divine love and grace out of nowhere.

SHELDON: Ha! In a BCC [Bible Chat Course] when I was a kid, the pastor once told us about the creation out of nothing. Isn't it remarkable that modern physics can explain this event absolutely scientifically?

THOMAS: Yes, the teaching of creation out of nothing has established itself only gradually in Christian theology. In its well-developed form, in fact, it evolved in the late Middle Ages. However, it opposes the hypothesis regarding the origin of the universe out of a quantum fluctuation. The primeval vacuum of your physics, with all of its symmetries and laws of quantum mechanics, is not the nothing from which God sovereignly called forth creation.[3] Your vacuum is already something in existence, a bubbling chaos full of possibilities, in which elementary particles spontaneously emerge and disappear again. In a historic perspective, this corresponds more to the primary matter of older cosmologies from which the cosmos surfaced.

SHELDON: So the modern conception corresponds more to the pre-Christian orient?

THOMAS: Not only the ancient Orient, but also the Greek cosmologies proceed from similar assumptions. For a long time, even the ancient Jews and Christians by no means ruled out a primary matter. When they began to interpret the creation narratives philosophically, they started to think about creation out of nothing, but without meaning a pure nothing. Besides Genesis, they also reverted to Plato, who talked about primary "space," a kind of pre-cosmic matter without actual existence.[4]

SHELDON: It's conceivable that in the primeval vacuum, all the physics of the later universe was already laid out. The creation of matter then took place according to these laws. But there was no matter and therefore no clocks or watches, and no yardsticks with which one could measure time and space or even define them. In physics one can talk sensibly about time and space only after the beginning of the universe. These basic elements of reality came about at the moment of the beginning.

THOMAS: The vacuum fluctuated already before the big bang. Gales stormed over the primeval ocean and the symmetries governed the to-and-fro of the waves! Then the coming and going of the waves' white crests already imply transient forms of space and time.

SHELDON: Physics can't say a thing about that, because absolutely nothing is left from these pre-cosmic fluctuations. There used to be speculations that some small black holes originated from the primeval vacuum and, under the influence of mutual gravity, they multiplied with a snowball effect. The time came when an instability began, opened space, and swept it along to a development of gigantic proportions.[5]

THOMAS: You seem to be shifting the actual origin more and more into the prehistory of the big bang. Therefore, creation out of nothing must be placed before the primeval vacuum.

SHELDON: No, it makes no sense to speak about "before the big bang." There was no "before" or "after" before time began. These kinds of speculations lead nowhere.

THOMAS: OK, but in theology, one likes to let time start with the beginning of creation.[6] It's no coincidence that the book of Genesis begins creation with the conception of light. Light brings with it the chronology of time, the change from day and night.

Back to the big bang and its consequences: I consider it significant that you assume the universe to have a beginning as well as a history.

SHELDON: Yes . . . There was a beginning. Look at the sky through the porthole. It's so black because the galaxies are not infinitely old. Farther away than about thirteen billion light-years there are no visible objects anymore, because the view falls back into the time shortly after the big bang, when no stars existed yet. If there had always been galaxies with shining suns, the night sky would be ablaze with light, similar to a light beam in a snowstorm that ultimately always falls on a white snowflake. The dark night sky shows, however, that the sphere of the galaxies is relatively small and that there was a beginning.

THOMAS: Further to the above, I recently downloaded a film about the *Post-modern Steady State Theory* (PSST) from our sciencetainment server. According to that theory, the universe is in an eternal balance. The new theory is based on the time-honored model by Bondi, Gold, and Hoyle, who in the late 1940s developed the idea of an infinitely old universe. It should be uniform in all directions, in space and time. According to this model, nothing changes on a large scale. New matter continually forms in intergalactic space and delivers replenishments for new galaxies and stars.

SHELDON: Of course, there are still such rather exotic attempts and perhaps we don't take them seriously enough. They remind us that all scientific statements are theories and could be wrong. Although observations will become more and more convincing, their interpretation is always subject to people who are sometimes mistaken.

Hoihong Wang presented a schema of the cosmological models discussed by the astronauts. All of the models originated from the late twentieth century.

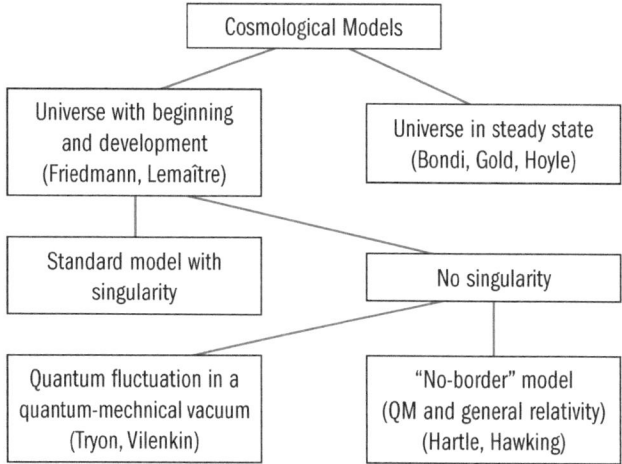

Figure 5. The schema brings the mentioned models from the second half of the twentieth century into a systematic order. It does not offer a complete listing of all modern world models, but sketches the starting points for all further developments since the turn of the twenty-first century.

THOMAS: Let's assume that the universe had an origin. Then the actual starting point naturally interests me. Here it seems that God's primordial work of creation almost becomes a physically describable event.

SHELDON: I'd prefer to consider first the *standard model* of the big bang theory. It's one of the many solutions that the general theory of relativity allows in space and time. This solution complies exactly with the expansion of the universe observed today. Therefore, it describes the universe in earlier times very well, provided that the physics then wasn't substantially different from today. The zero point of time, known as singularity, however, is a special case in this model. At that time, the universe would have been concentrated in an immensely small space. The density of matter was therefore infinitely large. No physics is possible anymore at this singular point of history of the universe, because the equations would have no finite solutions.

THOMAS: The singularity as an end of physics! That virtually calls for a metaphysical or even theological interpretation.[7]

SHELDON: Careful, there is a trap! I think such a singularity is totally unlikely. Einstein's equations of the standard model are certainly no longer valid. They were verified only in large systems of the size of stars. In the extreme situation of the big bang, the dimensions were so inconceivably small that quantum mechanical effects came into play, and the basic equations of the standard model don't apply anymore. Extrapolating into the singularity is an impermissible speculation.

THOMAS: Your knowledge can't go back arbitrarily, but stops a tiny fraction of a second before the big bang. I conclude that physics will never have the necessary language for the actual starting point and must capitulate.

SHELDON: Never say never! Models requiring no initial singularity have been suggested exactly because of such problems. I've already mentioned the theory of quantum fluctuation. At the end of the twentieth century, James Hartle and Steven Hawking undertook another attempt regarding the combination of quantum mechanics and general relativity theory.[8] They introduced an imaginary component of time and so bypassed the singularity of the beginning in their model.

THOMAS: Do you mean that time had no beginning?

SHELDON: The imaginary time has no border at the zero point but bends together with space so that they merge into each other. It's like with the polar region on Earth. If you continue straight north and cross the North Pole, you suddenly notice you're going south. Imagine you set your watch to decreasing imaginary time. Then you could go back to the big bang, walk through it—which would certainly be a bit painful—and you would come out with an increasing time. In the "no border" model, there is no singularity anymore and therefore no point where God would have to intervene. There are also no special boundary conditions necessary any longer. God is no longer needed.

THOMAS: This elimination of time is for me more offensive than the suppression of the creator God, who doesn't appear in all the other physical theories either. On the contrary, it would have quite surprised me if he suddenly turned up in your equations.

SHELDON: But just a short while ago, you used the big bang and its singularity to support your theological ideas.

THOMAS: On the contrary! Whoever speaks of the world as a creation refers to experiences of God today and doesn't refer to a long-ago hypothetical first moment. At that starting point, not only the equations of physics fail altogether but also the possibilities of human—and therefore also religious—imagination and language.

SHELDON: I'm still surprised how you disparage the meaning of this earliest moment. The first second of the universe has been for a long time a hot field of scientific research. Why have stories of creation had such a great importance for the religions? If it were only a question of purpose and well-being in the present life, there'd be no need for cosmologies.

THOMAS: Yes, the first moment of creation is also fascinating to me. Otherwise we wouldn't have discussed it for so long. Evidently there is an elementary connection between the primary event and the present experiences. Similarly, the interest of the physicists is not limited to the reconstruction of past incidents but addresses laws of nature that are still in force today.

SHELDON: The physics of the first millisecond of the universe has, in fact, become a key to the physics of the world's elementary components in general. Yet likewise, it's no longer about experiences in a day-to-day sense. The world of quarks, gluons, and strings is far more bizarre and foreign than anything we can detect through our porthole.

THOMAS: Contemporary experiences about creation are extrapolated to the origin in many religions. Human beings want to know where everything came from. Religious cosmologies want to integrate the elementary life experiences with this desire. One faces this task anew for every epoch. Earlier, we talked about religious perceptions of the cosmos that always also develop into a worldview.

SHELDON: Consequently, that would also have to be valid for the postulated creation-out-of-nothing.

THOMAS: In a religious view derived from the biblical tradition, new life constantly flows to us directly from God. In the cosmological horizon, this experience widens into the thought of God's uninterrupted world conservation. I'm aware of the *creatio continua* most clearly in the continuous creation of time. God constantly creates a new present. That's why the phenomenon of time captivates me so strongly. Time is not only a product of creation long ago but, quite simply, the act of creation as such.

"Creation out of nothing" received a significant role in the preceding discussion. For that reason, I'd like to show you a text from Professor Johannes Taubenschlag, professor of theology in Amsterdam, the Netherlands, for an explanation.

Creation Out of Nothing

The thought of the *creatio ex nihilo* (creation out of nothing) is regarded as one of the theological top principles of monotheistic religions, particularly of Christianity. The teaching means that the world as a whole with all it contains was called into existence and is maintained alone by God. With that, two important other conceptions are eliminated:

1. God did not create the world out of preexisting primary matter. The characteristics of the world are therefore derived totally from the divine will and not the additional "chaotic" effects of such an original matter.
2. The world did not originate as a consequence of God's self-unfolding, an emanation of his own being, but as his free will.

The thought of *creatio ex nihilo* is already laid out in the understanding of creation in the Old Testament's younger texts.[9] My colleague Thomas Haubensak points out correctly that the intellectual formulation of the doctrine first came about in the course of a long history. That the world is not necessarily as it is, but rather could look totally different, was made clear only with the nominalist theology of the late Middle Ages. One speaks in this connection of *contingency*. Contingent is that what can be, does not have to be. That is, a scope for the possible in contrast to the necessary. With this view, the composition of the world is not only contingent but also God's will itself. God could have not made the world at all or could have created it totally different. God and the world are clearly differentiated.

Two points attract attention in Thomas Haubensak's considerations.

First, it means a lot to him to interpret the *creatio continua*—the continuing conservation of the world by God—from the thought of *creatio ex nihilo*. That is, on the basis of a reflection on time. That is, to be sure, not unusual in Christian theology, as Paul and Luther demonstrate,[10] but could also be seen in connection with Thomas's interest in Asian religions.

Second, it seems that through his fascination with the "quantum ocean," he embarks on a real flirt with the "primary matter"—the chaos of the ancient Near East. That reminds me of the American "process theology" and doesn't get on well with the intention of creation out of nothing. In his defense, one should point out that even with highly esteemed experts such as Karl Barth, a peculiar interest in "nothingness" as a shadow realm "at the left side of God" can be noticed.

At the River of Time

SHELDON: After such a heavy discussion of creation out of nothing, I suggest, Thomas, that we reward ourselves with some of the very old Armagnac we have hidden away despite the opposing strict regulations of the space agency.

THOMAS: A brilliant notion, though you're right, we can't tell IASA how the drink got on this trip where every ounce costs hundred times that of gold.

SHELDON: Yes, when drinking just a sip of this Vieux Armagnac, I recall that while I toiled away with the air resistance device, your idea came to mind that time may not only be interpreted as a product of creation but also as an act of creation. When I look at our board clocks, however, their monotonous advance reminds me of everything else except a creative force—not to mention the fact that we are only arbitrarily synchronized with Earth time.

THOMAS: I too have been reflecting on the mysterious character of the flow of time. On Earth I lived for a long time in a city on a large river. We could swim in it during the summer. Its current was so strong that it was impossible to swim upstream. When I let myself drift along in the river, I often had to think about the time that carries us and sweeps us along. It pushes inexorably from the past into the future.

SHELDON: Whoever swims with the river locates the future downstream, where it's already waiting at the banks. Whoever takes a position outside the river and looks at it, time seems coming from upstream. You sound like you're talking about a time reversal. I do have some colleagues who speculate about the possibility of reversal in the direction of time. Some nostalgists even cling to the outdated model, in which the expansion of the universe would turn over into a contraction, ending in a collapse. Didn't Plato suspect that world history could run backward? Dead would rise from the grave, old men would become children, children would disappear into the womb.[1]

THOMAS: I didn't mean a purely formal reversal of the arrow of time. Rather, a qualitative difference in the perception of time. With your idea of time reversal, the future would be again the arrow's "where to" and the past accordingly its "from where." Time streams toward me like the river from the spring. It doesn't flow out of the past into the future but in reverse—out of the future into the past.[2]

SHELDON: How do you mean that?

THOMAS: Just take this moment when we enjoy the Vieux Armagnac. The present always becomes new. What is now becomes the past and a new "now" emerges out of the wide space of the future. The future is that which is coming toward me. I look upstream and from there comes the new.

SHELDON: I'll have another drink to that. So here comes another "now," and then it's gone.

THOMAS: That's it! Time flows here as well, but only in the opposite direction. The present always constitutes itself new out of the *not yet*. The present is then not a child of the past but of the future. It comes from an open expanse full of promises.

SHELDON: I suspect something: Would you like to catch the creator "in the act," there, where he busily creates time?

THOMAS: Ha! The advanced hour with its pleasant euphoria perhaps makes me a little imprudent. So I have to confess: Yes! It's still something deeply marvelous for me that out of nothing a new present will appear.[3] Normally we aren't aware of it because it's taken for granted. I identify the primary and most direct divine creation power in the becoming of the always-new present. For me, this is the most elementary expression of continuous creation.

SHELDON: If this continues, you'll delight us with a sermon. I remember a devout philosopher in our college who tried to inspire us for his course with the rhetorical question "Why is there something rather than nothing?"

THOMAS: But seriously, since a new present is always forming before our eyes, time gains an unfathomable—yes, even divine—dimension. The creative power is at work in the most trivial experience of our daily lives! One is probably confronted most intensively with this perception under extreme circumstances. Once, I was almost engulfed by a raging sea. When I staggered out of the water onto the beach, barely escaping drowning, I felt with every fiber of my body that life cannot be taken for granted—it's a wonderful gift. Since then, every second with its possibilities is a miracle for me.

SHELDON: Apparently you mean the moment, the immediate now. This moment is gone again instantly. You can't catch hold of God here either.

THOMAS: I don't want to.

SHELDON: OK, granted, at least for this late hour. Yet how do you get from this out-of-nothing-produced "now" to the reality shaped by causality and determination? Your "now" without origin is deeply formed by the respective past. Without our spacecraft built many years ago, we now wouldn't be happily floating in the onboard saloon having Saturn and Titan in view. Speaking with an image: your seemingly parentless child may be a foundling or orphan, but he has parents, naturally, who have given him precise genetic information to take with him.

THOMAS: Well, in one perspective, time flows out of the past into the future. All causal procedures derive from this time configuration and therefore all the forms of science known to us. In the other perspective, however, the future becomes the source of the present. Every moment breathes something from the freshness of the origin. It's the time experienced by poets and lovers.

SHELDON: Yes, but even the poets and lovers know what time will probably bring them. They make arrangements. The future is simply determined by the past, even if chance receives a significant meaning here. But for the digestion, we must permit ourselves another allowance of Armagnac!

THOMAS: We've already used up our weekly ration. The source has dried up. This isn't the right time anymore; the kairos—the moment to act—is no longer favorable.

SHELDON: As I'm already on your turf, allow me one other critical remark. You proceed from the assumption that newness portrays a kind of divine creation, and that the new even hides behind all that continuously exists. I can think of numerous examples where I find in this activity of God nothing more than sheer cruelty.

THOMAS: For example?

SHELDON: Deformities in newborn children. And these malformations are, according to you, preserved by God on a long-term basis.

THOMAS: You're asking the great old question about how God can allow something evil in his creation.

SHELDON: Well, you expressly provoked it . . . and the constantly diminishing number of your excellent Armagnac portions calls us for the night rest. So, for the moment I'll be satisfied if you at least give me the sight lines that you would use for your answer.

THOMAS: Two points are important for me. First, a lot depends on God's "empathy." When God takes up and preserves the existing in all of its ambivalence, it brings him also great suffering. The memorial of this empathy is the suffering Christ on the cross, when God exposed himself to suffering and decay.

SHELDON: Our discussions from the past Easter echo again. And what is the second point?

THOMAS: Here I try to perceive the signature of Good Friday and Easter as a fine structure of time in each of its moments. The resurrection of Jesus represents the new creation that each present brings with it. At the same time, this new must integrate into the limitations and physical laws that the previous was subject to. Thus it takes up the old. In that way, the risen Jesus has the scars of the crucified Christ.[4] He hasn't wiped off the past. The new doesn't simply eradicate the previous but preserves and integrates it. Seen that way, the cross represents the past preserved in faithfulness and compassion by God.

SHELDON: Evolution is a good housekeeper. What has stood the test will continue to be used or converted. I assume, though, that you are after far more than the banality of a good budget discipline.

THOMAS: I'm after a connection between the river of time and the resurrection of Christ. I remember a beautiful day in May when I was a young student and I became absorbed in a book on the Trinity of God. All of a sudden, the thought flipped through me: Today is spring because *he* has risen. Since then, I look for a way to understand this inspiration and generalize it. I remember the old church song "Christ Is Risen" dating back to pre-Reformation time. Its second stanza begins with, "If he had not risen, the world would have passed away."[5] Now you see how far Saturn, the star of the prophets and the philosophers, has already cast me under its spell.

SHELDON: I may accept these daring thoughts as a nightcap, since there's no time for any more of the delicious Armagnac and a fair amount of work awaits us tomorrow. The ODYSSEUS probe will reach the moons Enceladus and Tethys in a few days. Without our careful assistance, it won't have anything new to tell us! And now I have to take care of SMART. It just sent us its new position.

THOMAS: Understood. I'm inspired by our discussions about the beginning of time to write in my diary.

Cronus Alias Chronos
From Thomas's Diary (April 23)

Our last round of talks unintentionally covered topics that don't concern the classic philosophical problems but are surrounded with the nimbus of insolubility: the riddles of beginning and time. One bumps almost painfully into the borders not only of one's own limited intellect but also of conceptual thinking altogether.

Is it again our sojourn in the eerie cosmic border region that lures our thoughts onto unfathomable paths? The strange interpenetration of the incredible spectacular views in outer space and the monotony of being a hermit far away from earthly life stir up all the layers of my soul.

I must have dozed off while contemplating, when the old god Cronus visited me in my dreams again. It was not the charming beauty of the eternal spherical harmony but a terrible picture that overcame me in a nightmare—Goya's dark picture of Saturn wildly devouring his children. Grim recollections arise in me—the indications in the Old Testament about horrible offerings of children to the Moloch. It is no coincidence that Saturn in antiquity was also associated with Phoenician-Punic child sacrifices.[6]

In the ancient tradition, Saturn was promoted also to the lord of time! His Greek name Cronus was easy to associate with chronos *(Greek for time). The whole range of his mythological references could then be interpreted as a wide allegory of time.[7]*

Cronus is often shown with a sickle ready to castrate Uranus, his exceedingly procreating father. The sickle symbolized the merciless mowing that time continuously executes. It survived in the medieval image of the Grim Reaper, of death. Above all, it was a cruel trait of the old myth that won unexpected meaning by the naturalistic allegorical interpretation: Cronus swallowed all his children born by his wife Rhea right after their birth, out of fear of being overthrown by his own son.

Philosophy seized this atrocious myth in a subtle way: Behind the archaic drama of the Cronus story one found the figure of Chronos, time that immediately devours what it has produced. Chronos is at the same time father and destroyer of all who arose from him.

At the River of Time

Figure 6. Saturn devouring a child. His chariot is pulled by the dragons of time. (Aratus Solensis, Apud Theodorum Graminaeum: Cologne 1569. Copyright Bayerische Staatsbibliothek, Munich)

The myth also tells that Rhea was able to hide Zeus, the last-born son. Grown up, he forced Cronus to throw up all his siblings. The myth is symbolic for time: everything is swallowed up by Chronos, but everything returns back into the light. Not surprising, Cronus is connected also to fertility: like the earth takes up the seeds, Cronus devours his products, but some day they press back into existence.[8]

The molar of time! The voices of all those poets who complained in their moving lamentations about the inevitable decay that overshadows all life, echo in me. Tempus edax rerum! *[Time, devourer of all things!] as the Roman poet Ovid bemoans:*

O Time, thou great devourer, and thou, envious Age,
Together you destroy all things; and, slowly gnawing with your teeth,
You finally consume all things in lingering death![9]

The Birth of Time

SHELDON: Hey, Thomas! It's April 25, 2051. *Hermes Trismegistus* has completed its second orbit around Saturn and is still waiting in the Lagrange point between the planet and its largest moon, Titan. I've taken care of our faithful SMART, the unmanned Self-Maneuvering and Robotic Transporter. It has replenished the stocks of our colleagues in the ORPHEUS.

THOMAS: This occasion deserves a little celebration. We can have electric candles, an exquisite piece of tofu-made "moonfish" from the outside cabinet, and a bottle of the reconstituted Hungarian Bull's Blood wine that IASA explicitly allowed for lowering the cholesterol level.

SHELDON: Sounds delectable. And a good occasion for a further discussion about time and its characteristics.

THOMAS: Yes, I'm reminded of the opulent Saturnalia in ancient Rome. During the days of the winter solstice, the Romans imagined to have stepped back in time to the golden age of Saturn's reign and were feasting exuberantly. The master was slave, the servant became the master.

SHELDON: Well, we'd better not exchange our roles despite the Saturnalia. There is no way I'll let you touch the station control. Otherwise we'd be abruptly pushed back into the Iron Age that we're exposed to out here.

THOMAS: Don't worry. I'm fine with the station control being your domain entirely, and I'm content in my role. So this is the second time we have circled Saturn in our *Hermes Trismegistus*. I already feel part of this marvelous harmonious world with the moons as we roam our majestic orbit.

SHELDON: Have you noticed that the motion of the inner moons is already observable after one hour? We can compare our own movement to Iapetus, the third-largest and slowest moving of Saturn's moons.

THOMAS: The constant circling creates the sensation of an inner peace of mind. The symmetry of this motion seems to be perfect, and through it I feel a touch of eternity. I can almost hear the music of the spheres of Pythagoras.

SHELDON: Don't let yourself be seduced by the siren sounds of eternity! Time proceeds inexorably also in Saturn's system, even though the dominating movement for us seems to be cyclical. The mutual attraction of the moons builds up chaotic changes that seem to be random and are surely not timeless or reversible. With the contemplation of harmony, you move in a direction where time is finally downgraded to a geometric quantity.

THOMAS: At least I'm in good company. Isn't this geometric conception of time fundamental in the theory of relativity? Einstein's cosmic piety is virtually permeated by this harmonious view. He said that the religious belief of research scientists is founded on what he described as "a rapturous amazement at the harmony of natural law, which reveals an intelligence of such superiority that, compared with it, all the systematic thinking and acting of human beings is an utterly insignificant reflection."[1]

SHELDON: Yes, these words from a great man, which he wrote in January 1936, also touch me. Einstein's orientation to harmony proceeds from an all-permeating causality, which determines the future not less than the past. That was probably one of the reasons why he rejected the randomness of quantum mechanics. A world where chance opened the door only a few inches would have his "rapturous amazement" substantially curtailed.

THOMAS: Randomness seems to have no room in a world where every object runs its course like the moons around Saturn.

SHELDON: In view of geometric time, I agree. Time is only a parameter of the system that, for example, defines the location of a moon on its orbit. Let's assume Saturn's moon Rhea, gleaming so nicely in the sunlight over there, moves in an orbit with a known radius. Give me time and I can calculate the precise position of the moon. On the other hand, if you give me a location on the orbital course, the time can be given when the moon crosses the position. Time and space are firmly chained and equivalent, quasi exchangeable.

THOMAS: Let me develop the thought a little further. In this geometrical view, chance would be only possible when, for instance, a moon suddenly would have the choice between different courses—say, to continue or reverse. The continuation of the moon's course wouldn't be set by causality but purely accidentally. Anyway, for me, geometrical time is not congenial to space. They have a totally different content in your example. They aren't two of a kind.

SHELDON: That changes when the speed of light is finite. The clocks inside a moving system, such as a superfast spaceship, don't keep the same time from our perspective. Let's assume that two clocks—one in the front and one in the rear—show the same time for an astronaut in the cabin. For an outsider, the front clock is faster. Time depends on the location in the cabin and has therefore also a space-type character.

THOMAS: In my otherwise agonizing physics class, I experienced it as truly refreshing when our teacher elevated time in the rank of a fourth dimension. Everything became suddenly mysterious. I started to dream about fantastic time travels. I've never understood these assertions.

SHELDON: In the equations of special relativity there is almost no difference anymore between space and time coordinates. Time becomes part of geometry. The German mathematician Hermann Minkowski already established this in equations for the conversion of coordinates between moving systems. Einstein promoted these thoughts to a general principle about the invariable speed of light.

THOMAS: As fascinating as this concept may be, it opposes the human experiences and really all biological phenomena. Chance seems to have no room in the four-dimensional space-time, just as little as the merciless progress of time.

SHELDON: But chance and time cling together. In quantum mechanics, the veil is lifted by measurement through the observer. In that moment, a random event occurs that is irreversible. You can imagine quantum systems like the waves on a lake, but not as moving points. Here, time receives a completely different meaning: it brings forth reality.

THOMAS: No wonder Einstein couldn't accept that. With this view of reality, we get far from the everlasting harmony of spheres. Causality seems to make room for the purely arbitrary.

SHELDON: Quantum chance isn't arbitrary, because the probability can be calculated. By the way, I always give Einstein the right to criticize quantum mechanics. Anyone who isn't shocked by quantum mechanics hasn't understood it.[2] After all, it's a theory constructed by humans, and although tried and tested many times, it may still today possibly contain mistakes or limits. Einstein's objection that we may not know all parameters and some may still be hidden, however, is refuted by experiments. The randomness in quantum mechanics can't be explained away. Today we have to reply to Einstein's suggestive assertion, "God doesn't play dice" with a clear "Yes, he does."[3]

THOMAS: Your jump from physics into theology is too rash for me.

SHELDON: I thought you loved these artistic jumps!

THOMAS: No, I want to know more about the physical problem before we proceed to metaphysics. It's well known that Einstein refused to recognize quantum theory all his life. In this sense, he was the last significant representative of classical physics, and he seems to be absolutely right concerning our everyday world.

SHELDON: Analog problems face us in our mundane world as well. Chance isn't limited to the quantum world, but it is a part of all macrophysical processes. It's only a question of the length of time we're considering. Take, for instance, Saturn's moons. They orbit on almost perfect circles during the brief interval we visit them. Nevertheless, they mutually interfere with each other through gravity. And these small interferences mount up over thousands of years. Their orbiting positions therefore can't be calculated in advance over millions of years. We can, if everything turns out right, calculate their probable position, but where they actually wind up at any moment in the far future is chance for us.

THOMAS: You say chance "for us." We apparently have a knowledge deficit on the macroscopic level. It is not a fundamental indetermination. Chance doesn't really exist and isn't rooted ontologically in the being itself.

SHELDON: Yes, because we don't know the courses sufficiently on which the moons orbit. The uncertainty will become larger and larger. Since the error doesn't increase simply linearly with time, but shoots up massively, there will be a time when a moon's position can be anywhere on its circular orbit.

THOMAS: The impression of chance comes up only because we don't know the exact position at the beginning.

SHELDON: That's correct in principle, but in practice there are always limits to how exactly one can measure positions. With smaller and smaller dimensions, one arrives finally in the quantum world, where the positions are genuinely uncertain.

THOMAS: These two manifestations of chance differ substantially. In the macroscopic world, our limited cognitive capability opens the door to chance. In the end, chance is only apparent. In the quantum world, however, randomness is evidently not to be avoided. For Einstein, the second was far more offensive than the first. I wonder if the quantum world can let its dreadful state of affairs also play in the macro world. Can the elementary level below affect something on a higher level?

SHELDON: The nonpredictability of the course of the macroworld is very simply a result of the nonlinear equations of macrophysics. You're right, however, that the quantum mechanical uncertainty is more fundamental. The combination of both is known as nonlinear quantum physics, or "quantum chaos." It is one of the exciting fields for research in our century.

THOMAS: So, do we have to expect that the spontaneity or randomness of the quantum world affects our daily lives?

SHELDON: Certainly! Take, for example, a billiard ball. Every time it collides with another ball, the future course becomes more uncertain. I can easily calculate for you that after the eighth jolt, the initial indeterminacy at the quantum level builds up to such an extent that the uncertainty of the course becomes larger than the diameter of the billiard ball.

THOMAS: Einstein's unease was therefore well founded. If I understood you correctly, there are no indications that a kind of strict determinism still hides behind the accidental events in the quantum world.

SHELDON: No. The search for hidden parameters was unsuccessful. Of course, there are the ordinary laws of conservation that limit what can happen. Yet we basically do not know what will take place within that frame. I'll provoke you once more: God seems to play dice!

THOMAS: When faced with such assertions, our teachers emphatically warned us not to project the pattern of the intraworldly causality to the relationship of God and the world. Therefore, I hesitate to answer the famous question of Einstein with a simple yes or no. The problem is basically not new, although through quantum physics it has ratcheted up in appearance. In theology's history, the discussion about the relationship of divine predestination and human freedom has been important for a long time. It's about the connection between God and humankind. The projection of this specific relation onto events in nature is also problematic.

SHELDON: Wait a minute! You're evading my question and trying to escape into scholastic dialectic.

THOMAS: No, I don't want to dodge the issue. I'm only trying to understand what you really mean when you ask a question like "Does God play dice?" Both a yes and a no are shots in the dark.

SHELDON: Well, let yourself be content with a simple answer on this beautiful evening. Titan winks roguishly through the large porthole, and Saturn invites one to dream. And your stringent colleagues in theology are immeasurably far away.

Does God Play Dice?

THOMAS: The moonfish on the plate and the Bull's Blood wine lift my spirits. During the Saturnalia festival in ancient Rome, some people indulged excessively in the game of dice that was otherwise frowned upon. This gives me some freedom, too.

SHELDON: OK. I'll eat and you explain what you mean, if you can.

THOMAS: Let's assume that God doesn't guide the occurrences of the quantum world by virtue of his higher power. If I don't withdraw totally to the commonplace terms that God's ways are inscrutable and unknowable, I am forced to say yes to the question. God is playing dice! But—

SHELDON: Now it's getting exciting. But what?

THOMAS: I'll try to stay with the metaphor of playing dice, although I wince at the idea of God with a dice shaker in his hand! That means, first of all, even for God, chance is real, unforeseeable. But it also means God keeps to the rules of the game.

SHELDON: Allow me a short remark. The image of the game has become rather fundamental for modern science, because here the interplay of natural laws and chance can be studied in well-defined examples and can be simulated.[1] The world is an open game, only the rules are set. There's lots of room for chance. At the same time, it's channeled by laws.

THOMAS: The metaphor of a game also has a home in biblical tradition. About the divine wisdom that participated in the creational work, it's said that wisdom played devotedly before God.[2]

SHELDON: So is God a gambler?

THOMAS: God takes part in the world game in any case—whether there are other players, I won't even ask. Speak of the devil and the devil shows up! I suggest that God himself laid down the game rules.

SHELDON: Here you collide unavoidably with evolutionary theories. Many laws in biology became effective only with evolution. Laws themselves have a natural history.

THOMAS: I see God's work also in processes described by evolution. Furthermore, I learned from you that many basic laws, such as symmetries, already were valid at the beginning and, so to speak, watched over the pre-cosmic quantum ocean.

SHELDON: Granted. But God must reckon with chance anyway.

THOMAS: With chance just as much as with the once-designated laws. Now I have to add something else: God keeps the whole world game running. We've talked earlier about the continuing preservation of the world. He is the host who invites us to the game, and he is at the same time a player. He has the possibility to stop the game at any time.

SHELDON: Then he would be a spoilsport!

THOMAS: Yes indeed! Many biblical texts speak about God's anger or indignation regarding his people in reaction to a rather adverse play that human history takes. Yet he doesn't stop the game. More likely, humans are the spoilsports.

SHELDON: I thought that God's omnipotence is hardly the question anymore.

THOMAS: That depends on what we mean by the term *omnipotence*. We should talk about it more in detail. For the moment, I'd like to stay with the imagery of the game.

SHELDON: Go on . . .

THOMAS: God keeps the world game running, even where it doesn't develop according to the way he wants. In this I see his faithfulness toward his creation but also his "empathy." He is connected with his work to the highest degree.

SHELDON: But he doesn't have any other possibility than to submit to the play. Otherwise the game would end in chaos.

THOMAS: It's not only associated with delight and satisfaction for him, but also with grief and sorrow. At this point, the colorful play changes to deadly serious. God exposes himself to all the chances of an unforeseen circumstance.

SHELDON: According to your view, he plays a double role: he is the preserving creator and, at the same time, the affected player. Let me speculate a little, stimulated by some Bull's Blood wine: I would talk about divine participation in the game where evolution is advanced through decisive and irreversible events.

THOMAS: You remind me of our discussion regarding Easter and the beginning of something new. But the metaphor also has a drawback. God is not only on the side of the successful. As a player, he also sides with the loser. In the biblical tradition, he lives with the deprived people, the poor, the maltreated, where life didn't butter the bread on both sides.

SHELDON: Now you've gone too far with this image of the game. We want to debate about the universe, not about social welfare and care for the poor.

THOMAS: Perhaps I jumped too far. But we've touched on the dark side of the play metaphor. Next to winners, a game also must have losers. Games can turn deathly serious. We can't mask all this out when we discuss Einstein's question.

SHELDON: Einstein's irritation about the divine dice player has a totally different background. He had a question regarding the steadfast laws of the universe that stimulated his admiration. He appreciated the divine in its paramount wisdom and beauty. I can understand him. I have more difficulty in connecting religious feelings with a dice-playing God. From my view, nothing holy clings to chance.

THOMAS: Religions offers many different interpretations. You've already mentioned one type: the universe's marvelous laws lead people to astonishment and reverence. Perhaps you know the verse in the Psalm of David in the Old Testament from Haydn's *Creation*. Let me sing it to you.

> *The heavens are telling the glory of God;*
> *and the firmament proclaims his handiwork.*[3]

I'll say more. In the Western world, the contemplation of the harmony and beauty in the cosmos played an important role for the concept of God since the earliest Greek philosophy. Einstein is in a venerable tradition of cosmic piety. The cognition of divine laws also made possible a correspondingly appropriate life complying with the universe.

SHELDON: In this perspective, space has apparently a dominant position. Plato said he held geometry for the path to God.[4] In antiquity, it was easier to make space the nexus for religious sentiments. Today, space is no longer ordered hierarchically; it is rather isotropic, equal in all directions. Also, the fundamental symmetries of the universe don't tell us any more about the glory of God. Let's leave Einstein's perception as it is. What about the counterpoint of space, about time? Is there a preference for time in some religions?

THOMAS: The type of cosmic, *philosophically* tinted religiosity we discussed has an extremely distanced relationship with time. Basically, there is only the eternal present. It applies largely to the Greek philosophers, and if I remember correctly, a similar idea comes up again, not coincidentally, with Einstein.

SHELDON: Yes, I remember it precisely. I even recited his words at a funeral oration. When in his old age he received the news about the death of a close friend, he wrote to the widow: "Now he has departed a little ahead of me from this quaint world. This means nothing. For us faithful physicists, the separation between past, present, and future has only the meaning of an illusion, though a persistent one."[5]

THOMAS: There are many indications that the oblivion of time in antique metaphysics has been preserved undiminished in today's physics.[6] A striking example is the continuing search for timeless laws of nature, an order that does not change and that is always valid. In the realm of *religions*, there are different opinions. Consider, for instance, the importance of the calendar and the holy times that even here in our isolation have given life an elementary rhythm. The cyclic course of the seasons provoked religious interpretations. Many old cultures revolve around growth and fertility. Knowledge of the order of creation and conduct of one's life were interlocked with each other.

SHELDON: That sounds like a time experience that's closer to reversible time.

THOMAS: That's correct, but I don't know much about natural religions. Time is constantly renewed in cultures close to nature. The more advanced civilizations in the ancient Orient cultivated and later on reproduced mythical creation events in great annual feasts. At the same time, they became aware of their own historicity.[7]

SHELDON: That means the advancement of time is noticed.

THOMAS: One could put it that way. In the large monotheistic religions, this conception of linearly advancing time becomes dominant. God acts between initial creation and eschatological completion. History itself becomes the preferred field for the revelation of God. God creates continuously new things that weren't here before.

SHELDON: That's why you're interested in time and irreversibility! This discovery of history has actually marked modern biology and astrophysics deeply.

THOMAS: Yes, but at the same time, the mythical margins of primeval time and the last days have evaporated. The modern age is at the mercy of history without prospect of a safe ending. It feels the horror of history without protection.

SHELDON: We'll have to discuss end-time scenarios another time. Now I've lost sight of chance, our alleged main subject. What does it have to do with religion and God?

THOMAS: Chance is a troublemaker in that type of cosmic religion that still echoes with Einstein. In ancient times, it wreaks havoc in the lowest cosmic sphere, the sub-lunar world. The reason is in the irrationality of human matter. Perfect regularity prevails, however, in the region of stars. The orderly movements of the stars express the divine harmony.

SHELDON: The movements of the planets were a prime example for the success of classical physics at the time of Newton. However, we know today that their apparent stability is only an idealized approximation. In cosmic timescales, their orbits are variable and even chaotic. Chance is here also at the helm.

THOMAS: In the religious perspective, coincidence is differently perceived. Let me play a little with that word. A coincidence is an incident that happens to people quite unforeseen and unexpected. To recognize a divine influence in unexpected incidents is certainly one of the most elementary forms of religious experiences. Some time ago we talked about kairos, the opportune moment. In the Greek world, a trouvaille, or a lucky find, was considered a "gift from Hermes," a *Hermaion*. Hermes, the winged messenger of the gods, who also helped the travelers, merchants, and even thieves, was considered the origin of such lucky changes and surprising events.

SHELDON: Let's drink to the goodwill of Hermes! May our *Hermes Trismegistus* bring us safely back to Earth.

THOMAS: Many parables from Jesus about unexpected happy events are conveyed in the Bible.[8] A man discovers a treasure in a field. A merchant stumbles on an outstandingly fine pearl. A king unexpectedly exempts all debts from a debtor. A vineyard owner surprises his workers with his unconventional payment of wages. A caretaker boldly changes the promissory notes. A banquet takes place with spontaneously rounded-up replacement guests. For these people, it is as if something surprisingly falls from heaven. The fog breaks up and a shining light flashes. The love of God surprises the world.

SHELDON: You should name here also the numerous stories of miracles in the Bible. Blind people became sighted, lame people could walk. But I've noticed that you have mentioned only pleasant surprises. Things change for the worse only too often. Think about the failure of the laser diode.

THOMAS: Yes, of course there's also the downside. A well-known parable warns about a thief who comes at night. Texts announcing the Last Judgment herald a threatening turn of events that will thwart all human calculations. A self-satisfied rich farmer suddenly dies. The divine judgment comes as unexpectedly upon the world as the Flood.[9]

SHELDON: I hardly know the Bible and can't fill in your short hints with life. Isn't it that chance apparently surprises people, but you're saying now it's used and guided by God? The question remains whether there's something accidental for God himself—a development of incidents that he couldn't foresee.

THOMAS: You've introduced a topic that leads to unresolvable logical contradictions, like the Cretan who says all Cretans are liars. For the moment, I'd like to suggest only one point where the problem crystallizes especially clear. In theology, one always wondered whether the human free will sets a significant limit for the divine power.

SHELDON: Have your authorities found ingenious answers for this thorny issue? The modern challenge is that the *cosmic* power of chance has entered the limelight as never before. This starts in the quantum world and recurs in molecular and biological evolution. It's about much more than only humans.

THOMAS: I can't give you a conclusive answer regarding the relation of God and chance. We're agreed that surprising things happen in the world time and again. The Old Testament even dares to talk about God's remorse about his own decisions. God so regretted having created humans, including all creatures, that he brought the Flood over the earth. After that, he supposedly decided to never again totally destroy the world. What a memorable learning process by God![10]

SHELDON: Indeed. And we shouldn't regret this nice evening. Tomorrow I have to contact the ORPHEUS crew very early. We're supposed to correct their mapping data regarding changeable surface conditions and integrate them into a map spanning the whole moon. We can contribute something important for the landing with this task.

THOMAS: I'll take this opportunity to write in my diary.

Between Space and Time
From Thomas's Diary (April 29)

I enjoyed the festive mood when we celebrated our second orbit around Saturn, but I'm caught between the poles of ambivalent opposite feelings and thoughts. Two forces wrestle in me and each of them seems to be the larger and more powerful: space and time.

I imagine a fictitious debate: the Macedonian king Alexander the Great lets a group of sunburned wise men in faraway India argue about who is more powerful, space or time. A very old man argues, "Time, because it is older and at the same time younger than all the others." Another person—a young man—opposes him: "It's space, because time would have nothing to devour without space." The king declares the young man as the winner. He himself—Alexander—conquered the entire world in little time and thus helped space to victory over time.

When the young ruler died only a few years later, however, an Indian envoy said, "Oh dear, great Alexander, you were so powerful over so vast a space, but you could not prolong your life even for a minuscule period of time. Now you must make do with a small grave, and time will soon also divide and crush your empire."

Father Space and Mother Time

A double image comes up in my mind's eye: the ocean *and the* river. *Both images individually have frequently accompanied me in my evening contemplation. Now they appear together, an unequal pair. Father Space and Mother Time!*

The ocean—endlessness of space! On its surface, a continuous surging, an up-and-down, back-and-forth, a coming-and-going. And time in its rhythms creates only a surface form of space. In the depth of the sea, however, is silence. There is no time anymore; there is sheer eternity. And me, a drop from this large sea, I feel driven by an ardent desire to return back into the One, to merge with it.

But there is also the river, the irreversible stream of time! It carries me away into an open vastness. All that is past is safe and secure in it. I feel a tremendously attractive power from the future that affects the stream.

Both ocean and river are water. What is the one *element from which space and time are made?*

Eternal Noon

Saturn's rings, together with their numerous moons, stir in me the intensive cosmic feeling of a grand majestic expanse. The old idea of the cosmic spheres revives in me: Blissful orbits around the central body. Pythagoras's music of the spheres and Dante's fiery heavens with glorifying hosts of angels. Enraptured view and ecstasy. One cosmos, resting in itself and suffused with one *great harmony.*

In the middle of this cosmic round dance rests the deity herself—origin of all, wrapped in unapproachable light. Everything comes from her and goes back to her. Eternal present, eternal midday! Time—a daughter of space, engulfed by the spheres, revolving forever ... When modern physicists talk about time as the fourth dimension, they still pay tribute to the reign of space over all temporal. People's deep longing: to change time into space, to freeze the ceaseless flow to eternal present.

Then there is the other global power—time. The orbits of the moons around Saturn, the orbits of the planets around the sun—in cosmic chronology, they are just episodes in a much larger course of time. The spherical harmony is only a short beautiful spring day in a vast chaotic world year. Even the present orbits are overlaid by totally noncyclic movements in the great universe. Admittedly, our sun may still circle around the heart of the Milky Way, but our galaxy flies on a direct collision course with the neighboring Andromeda galaxy. The whole group of our galaxies is pulled by far more powerful complexes of galaxies, whose names arouse my mythological fantasy; the galaxy supercluster of Virgo and finally the enormous supercluster of Laniakea far away in space with a frightening accumulation of gravity at its center: the "Great Attractor." There is no orbiting anymore, but a unique, perhaps straying course through the yawning void of the universe.

Finally, there is the expansion of the universe. Space inflates out in all directions with colossal speed. The universe has a beginning and maybe also an end; at least it has a unique history with specific eras and world ages. Space submits here again under the reign of time.

Religious feelings of a totally different kind are ignited by this cosmic history. It is no longer an entranced orbit of timeless celestial bodies and blissful spirits, but a start into an open vastness, an exodus from the old into the new. The recollection of one of the most magnificent biblical texts comes to me.[11] *When Moses wanted to see God's glorious splendor at Sinai, this ecstatic blessed view was denied him: "You cannot see my face; for no one shall see me and live."*

So Moses moved into the protection of a cleft in the rock and looked as God passed by. He cast a glance at the back of God, on his train full of light and power: "And you shall see my back; but my face shall not be seen."[12] *I interpret this passing of God as God's march through the times. God shares with his creatures the start from the old into the new; he walks ahead of his creation into a mutual future. His face remains hidden from humankind. He allows himself only to be found and recognized after the event, after his "passing by"—until he unveils his face of gleaming light at the completion.*

My mental games grope still further. I find all those histories from which the universe is made at "God's back." God's face remains withdrawn from us, but in time—in this living fabric from which the world is created—we can see and contemplate his work.

Eternity—Space or Time?

Between space and time there is a strange connection. When I look up into the sky, my gaze goes deep into the past. The stars' light may be years or millions of years old. Distant space is also bygone time. I ponder the paradoxes of relativity, even though its mathematical explanation remains incomprehensible to me. And there are scraps of memory of pithy sentences from a philosophical physicist that accompany me on my adventurous trip to Saturn: "The growth of space is . . . the openness of the future."[13] I wrestle with its meaning. If time is most fundamental, then space must be a manifestation of time—a quasi-frozen temporal configuration.

Now what about eternity? Is space finally winning back the reign over time? Of course, theological tradition has sharply differentiated between temporal infinity and God's eternity, wherein past, present, and future are not ordered one after the other but arranged "at the same time" and "into each other."[14] Here, the basic signature of our experience with time perception, the difference of "before" and "after," is masked; their irreversibility disappears. Again, space triumphs with its coexistence!

Can a thought about eternity be formed at all that does not betray our time perception at the end? I stretch out toward the blissful view, for a final and highest harmony of space and time, unveiled as if in a dark mirror. I feel dizzy. The concepts disintegrate in the middle of my musings. Again I bump into insurmountable limits. The blood pulses in my head; the fireworks of my neurons threaten to get out of my control. Have the first messengers of lunacy reached me that distress us earthly creatures from the infinity of the universe?

SHELDON: Bad news. There's been a delay in the landing on Titan of our five fellow astronauts and it's apparently beginning to cause them some psychological stress.

THOMAS: But they've been hardened in handling unforeseeable and adverse circumstances.

SHELDON: Right. Evidently the suggestions of ground control to relieve the tension by means of ergotherapy and overtone singing produced more the opposite effect.

THOMAS: Ha, I'm not surprised.

SHELDON: I'm going to use this delay to write another report for the Cyber School.

A Model of Reality
A Postcard from Saturn for the Cyber School

Hello, Earth!
You may have wondered why there haven't been any updates about landing on Titan. The reason is there have been difficulties in finding a suitable landing place on Titan's surface, which is furrowed with meteorite impacts, and methane bogs, rivers, and glaciers.

For weeks, our colleagues on the orbiting Titan landing craft ORPHEUS *have been using three radar units to map the surface of the moon, which is totally hidden behind photochemical mist. One of the radars works like a sonar and measures the height of the mountains and craters, while the other two screen the surface at a flat angle to determine their roughness.*[15]

Two problems came up simultaneously. The single measurements give different readings at the same location. The different radar units not only register other values, but also the observations with the same instrument differ at various times because of the limited measuring accuracy and disturbances in Titan's atmosphere. The second problem is the scarce scanning of the surface with these measurements. It is structured on such a small scale that we need a great number of measuring data in order to make a dependable map.

The computer on the landing craft ORPHEUS *takes the measuring data and calculates a model of Titan's surface. The more measurements gathered, the better the map will be. The approach is typical for science. The algorithm precedes from a starting model—in our case, the surface of a sphere. The computer then calculates what the outcome of measuring should be and compares the result with the measured values. From the deviation the computer constructs a better model with less variance. Because the measurements scatter and contradict each other, the discrepancy is never zero. The best model is that whose sum of all discrepancies is minimal. Because new data are added daily, the model improves in the course of time.*

Our strategy is to wait until the model doesn't change substantially in spite of new data. We assume that the models converge toward reality and finally conform to it.

Our models converge very slowly, and we don't know why. The reason is still puzzling us. New models show strange new structures on former flat areas we had already considered for landing. The surface may change because of precipitation that deposits yard-thick polymer mud.

From Saturn with best regards, Sheldon Cutter

Models—Constructs or Proxies for the Truth?

SHELDON: Our model of Titan's surface is finally converging in the last few days. It seems that some unexpected variability has delayed the solution. Now the more we observe the same location, the more accurate the mean value becomes and the model converges satisfactorily. The fog has lifted, Thomas! We are close to the truth.

THOMAS: Don't let yourself be fooled. A model is never reality. A computer on the Titan landing craft produced it from an artificial program. The model can be wrong because it is, like any theory in the end, a human construct. It imitates perhaps the truth, but, by far, it's not the truth itself.

SHELDON: I'm not so easily put off. Our colleagues are risking their necks by trusting that model. You'll see it will correspond with the truth.

THOMAS: The truth can't be equated with models and theories. I suspect that scientists are secretly fundamentalists on this point. The adepts have blind faith in a theory such as quantum mechanics that has become conventional. You let yourselves be more strongly impressed by the authority of age than you'd like to believe.

SHELDON: Not so! It's the great dream of every physics student to find a test result that contradicts a time-honored theory. Imagine that two particles collide and don't behave according to quantum mechanics. It could mean a Nobel Prize! Granted, a theory that has been proven so often won't be tested every day. That's not because of the age but the minuscule chance given to such an experiment. We trust in these theories and so do you, Thomas. You board a spaceship such as ours because you know exactly that the laws according to which it was built are reliable. A proven model can't be differentiated from the truth on the practical level.

THOMAS: But let's take the example of models from biology or theories about the early universe. Didn't some of them result from religious or ideological worldviews? In the late 1920s, a catholic priest, Georges Lemaître, proposed the first big bang model of the universe. Because of his profession, it was mistrusted by others who preferred a universe without a beginning. In the second half of the twentieth century, unstably growing processes became popular in science. They helped to make the big bang and cosmic inflation theory acceptable. At the end of the twentieth century, catastrophe and chaos theories were popular. Here again is the palpable influence of worldviews and the history of ideas. People construct convenient theories in every age.

SHELDON: Yes, worldviews have often stimulated theories, but only the initial models. Then they're compared with new observations and the theory has to be changed accordingly. This happened to cosmic expansion in the 1960s. Finally, the theories converge and change slowly or not at all. This definitive theory is independent from the initial model and closer to the truth.

THOMAS: The intensity with which biological theories and models are discussed and the longevity of alternative theories tell a different story.

SHELDON: It's only a matter of time until a theory is as well established as, for instance, the periodic table of chemical elements. Eventually it will also be the case with the theory of evolution.

THOMAS: The scientific method is reductive and selective. In order to explain certain observations and measurements, others are excluded. For example, esthetic and religious experiences connected with emotions won't be considered. There's no place for them in natural scientific models. How can one come to a definitive model with such simplification?

SHELDON: On the contrary, refraining from subjective impulses leads to more objective theories. My personal sentiments have no place in a scientific model.

THOMAS: But your cognition, Sheldon, is due to a specific perspective that is connected with the historical location of you, as the observer. Every science that doesn't account for the blind spot of its perceptions gets a distorted image of reality. What you essentially block out at the beginning assuredly won't show up any further in the cognitive process. The wide area of nonobjective perceptions that require participation of the observer is totally absent in your understanding of cognition. They not only concern phenomena that are essential to understand religion but also, for instance, to experience art.

SHELDON: We already discussed that at Easter, and I still don't get what you mean by "participatory perceptions." In physics, we know particles and wave fields that convey information. What else can we detect with our senses or instruments? Everything else is subjective interpretation by people and has nothing to do with reality.

THOMAS: You seem to believe that science reduces human subjectivity to a disturbing factor. It reminds me of the infamous "theory of everything" from the 1980s, when theoretical physicists in all seriousness claimed that they were close to the super theory that could explain everything in the universe.[1]

SHELDON: That was the attempt to integrate all four basic forces of physics into a common unified theory. Because the forces determine the dynamics of particles, the theoreticians believed they were close to the fundamental solution of all the world's riddles. I agree that physicists are sometimes overly optimistic. Most physicists at the end of the nineteenth century thought they were within reach of the completion of physics. But only a few years later, Max Planck began to study the radiation of black bodies and with that laid the foundation for the quantum theory, a totally different physics.

THOMAS: In the past, it was the philosophers and theologians who claimed to have a supertheory that can embrace everything. It seems that, today, theoretical physicists have devoted themselves to this program. They are the true heirs of the ancient European metaphysics! It is not the theory as such that irritates me but the incredible arrogance that accompanies this narrow perspective.

SHELDON: The "theory of everything" was undoubtedly an arrogant name. Even if the theory had been successful, it could only explain elementary particle physics. This branch is certainly important, but it doesn't contain the complete physics, by far. On the other hand, your inclusion of subjective perceptions in the process of cognition is still suspicious to me. I don't want to fall back behind the achievements of the Enlightenment. I love clarity and detest everything dark and obscure. I am quite ready to open my perspective to a reality that isn't embraced by scientific methods if somebody can convince me about the reality of "participatory perceptions."

THOMAS: But we can't continue this discussion now. Larger quantities of food, oxygen, and other items have to be sent to the Titan landing crew on the SMART transportation robot.

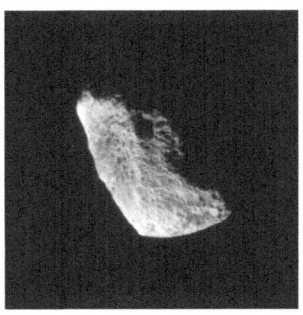

Figure 7. Hyperion has a nonspherical shape and is presumably a piece of debris from a cosmic collision. The moon is in resonance with Titan and Saturn, which makes its rotation and course chaotic. (Photo: NASA/JPL-Caltech/Space Science Institute)

In the Midst of Chaos
A Postcard from Saturn for the Cyber School

Hello, Earth!
We are closely watching the formation flight of the ODYSSEUS *probe with Saturn's strange moon Hyperion.*[2] *The little moon lurches like a drunkard back and forth and constantly changes its axis of rotation. Its revolution period is different with each orbit. The rotation is* chaotic, *as we say*

in physics. The gyration is only weakly stable and reacts heavily to the smallest outside influence.

What causes the effect is that when Titan makes three orbits, Hyperion circles exactly four times. The two are in resonance. The mighty Titan has a firm control on Hyperion. Titan forces it into a strongly eccentric elliptical orbit. Hyperion wants to adjust its own rotation to Titan's orbit, but small disturbances of Titan's orbit by other moons bring the midget to stagger.

Chaos means that no long-term predictions are possible. At our mission's takeoff three years ago, Hyperion's present location couldn't be calculated even with the fastest computers. Hyperion tumbles and sways as if there was no order in the solar system, no firm law of cause and effect. Therefore, ODYSSEUS *keeps a proper distance.*

Hyperion's wobbling motion makes it an apparent misfit among the pearls of Saturn's system. However, if one looks a little longer—I mean, some million years—all moons behave chaotically. Their orbits mutually interfere with one another and the uncertainties build up until they are getting the upper hand. The position of the moons cannot be predicted in the long run. Resonances can even change their orbits so there could be collisions between moons. It doesn't mean that their orbits necessarily end up in catastrophes. Their future is simply unknown.

For how long can we calculate the future? It all depends on the system. For large bodies, the orbit remains nearly constant, but the orbital position gradually changes in a chaotic way. For the inner planets such as Earth,[3] *it is one hundred million years; for the large Saturn moons, some ten thousand years; and for Hyperion, only a few months. We know the laws and forces of motion of these celestial bodies thoroughly, but we aren't able to make long-term predictions.*

The tumbling and swaying of Hyperion has a second reason. It's not spherical. Rather, it's twice as long as it is wide, similar to a potato. So it differs astonishingly from its sister moon, Mimas, also out of ice but almost a perfectly round globe. It must have something to do with the unique history of Hyperion. Maybe it's a fragment of a larger moon that broke up in a collision.

An eighty-mile-wide crater named Herschel blemishes the small Saturn moon Mimas. The landmark has a third of the moon's diameter and can be seen even with the naked eye from the Hermes Trismegistus. *An object with a diameter of approximately five miles must have smashed into the moon a long time ago, almost blowing it apart. If the violent visitor had been a little larger, Mimas would not exist anymore, or perhaps it would look like Hyperion.*

An unpredictable disorderly behavior looms behind the timeless façade of Saturn's charming system. The staggering Hyperion is the odd one out of the family, whose members are also basically misfitting. Its reeling reveals the clan. Greek mythology has already related stories about catastrophes from the sunken world of Cronus and the Titans. It also knows dark prophecies about the collapse of the currently ruling Olympian race of gods.

From Saturn with best regards, Sheldon Cutter

Figure 8. The Saturn moon Mimas is striking by its relatively large crater, named after the German-British astronomer [William] Herschel. The crater measures eighty-seven miles, almost a third of the moon's diameter. The shock wave of the comet or asteroid that caused the crater even left traces on the other side. The impact almost destroyed the little moon. (Photo: NASA/JPL-Caltech/Space Science Institute)

Chaotic Prospects

SHELDON: I hope that we'll soon know enough about the ground structures so the Titanauts can land safely. A residual risk remains in any case, since we can't calculate everything. Just look at Hyperion's stagger.

THOMAS: I still feel fairly safe in our orbit that, according to your remarks, is certainly also chaotic. In any case, we have reached the target point in Saturn's system well and precisely.

SHELDON: Actually, *Hermes Trismegistus*'s orbit is also chaotic because we can't calculate where it will be in a hundred years. For shorter time intervals, however, very exact predictions are possible. That's why we have a well-planned yearlong trip behind us, without noticing any chaotic incidents. The calculation will become imprecise and finally impossible only for larger time spans.

THOMAS: According to that, chaos wouldn't be noticeable within human lifetimes?

SHELDON: It depends on the system and its time constants. With the weather on Earth or Saturn, the forecast is well known to be very difficult beyond one week. The multitude of equations with different interacting masses of gas working together makes a long-term prognosis impossible.

THOMAS: Indeed, the future is wide open. That is an old life experience. Changes from one second to the next are, in fact, very seldom. The larger the time span, the more uncertain the future. In the long run, the universe is evidently tumbling into chaos.

SHELDON: That's actually the case. Take, for instance, Earth's orbit around the sun. It was described by Copernicus as a perfect circle, but then by Kepler as an ellipse. Today we know that both are only approximations. Earth's orbit is also chaotic. Earth's position is known now by better than six inches. But we're not in the position to calculate where our home planet will be in its orbit around the sun after one hundred million years. Small interferences by other planets will build up so much with time that the inaccuracy of the position after one hundred million years will be as large as Earth's orbit.

THOMAS: If one could measure the position of Earth more exactly, the prediction time would be prolonged.

SHELDON: For a two-hundred-million-year prediction, the center of Earth would have to be much more exactly known. Not only about the half—that is, three inches—but rather one angstrom, the diameter of an atom. That is the difference between a linear system, in which the error increases proportionally to the time, and a chaotic system, whose error grows exponentially.

THOMAS: It makes me think that even the apparently regular movements of the planets should be chaotic. Humankind's ancient cultures saw divine powers in the planets, today still witnessed by their names. Their stable motions represented for the Greeks the immutability of the gods and their exemption from the time. Saturn marked the limit between the world of planets and the highest divine sphere, that of the fixed stars. Physics and theology as metaphysics were here almost identical. But now modern science totally destroyed this view.

SHELDON: I don't regret it. The old physical model of the universe was like clockwork, predictable in all the future and the past. Accurate prediction would have given unlimited power to science over nature. Don't forget that at the starting point of modern science with Francis Bacon, about 1600, was the urge for control over nature. Chaotic processes were of no interest at that time. The first hints of chaos were recognized only in the nineteenth century, but even then, the attention was aimed at what was technically useful and granted power.

Chaotic Prospects

THOMAS: Then only the discovery of chaos could have broken the human presumptuousness.

SHELDON: The detection of chaos was one of the great scientific developments of the twentieth century. It's the liberation from the notion that everything, even human beings, is machinelike. In the chaotic universe, there's space for the unexpected and new. Remembering our discussion about the divine creation of something new, I wonder if this modern scientific view keeps a space free for God.

THOMAS: And I wonder whether chaos not only limits humankind but also sets limits for God. You told me that chaos is not a technical but an intrinsic problem. A computational limit of prediction will always appear at some point.

SHELDON: Chaos is definitely not simply an engineering problem to be solved by a more powerful computer or by improved accuracy of the measurements. On the mathematical level, the principal difference between linear and chaotic systems is definite. Linear equations usually have a solution, chaotic ones don't—their solution can only be approximated numerically.

THOMAS: That makes sense. But can you explain a bit more?

SHELDON: Up to now, chaos theory has been oriented on classical physics. Eventually, though, the uncertainty of quantum mechanics makes itself noticeable.[1] In fact, here one runs into the fundamental boundary that you asked about. The quantum mechanical uncertainty prohibits even Laplace's demon to tell the long-term future.

THOMAS: This strange demon has always intrigued, fascinated, and perplexed me. What do you actually mean by that?

SHELDON: It's just a thought experiment. Pierre-Simon Laplace published the first article on causal or scientific determinism in 1814. His basic idea was that the past completely determines the future. If someone (the demon) knows the precise location and momentum of every atom in the universe, their past and future value for any given time can be calculated from the laws of classical mechanics. So it only needs to insert this information into the laws of dynamics in order to obtain a world model for all objects in the universe, from the lightest atom to the largest celestial body. Past and present are equally known to it.

THOMAS: Then this demon is also a figure of theology. According to classical Christian tradition, the past and the future are present for God: all modes of time are simultaneous. In modern terms, it means that he knows the status of each particle and energy quantum in the universe from the big bang until the farthest future. That's why God is omniscient, because he sees all things to come as if they were already present. I'm all the more curious to know what happened to Laplace's spirit in the further course in the history of science. It hardly foresaw it would be dethroned!

SHELDON: The modern chaos theory didn't actually dethrone it, since, as I said before, chaos still moves to a great extent on the path of classical physics. The demon sees sharply in both time directions. It knows the totally unpredictable occurrences far in the future as well as the initial conditions of the chaos processes that we can't reconstruct. It doesn't matter for our thought experiment that the demon's database needed far more atoms for storing this immense knowledge than the universe actually contains. It's only quantum mechanical uncertainty that limits the demon's all-knowing.

I'd like to ask again whether this has any consequences for your understanding of God. Doesn't science's newly discovered openness again grant God's future a meaningful place? Our former discussions led us far in this direction.

THOMAS: God has certainly been freed out of the atrocious coffin of the nineteenth century's world machine. Nevertheless, the latest scientific findings can make some old images of God falter. You ask what the venerable statements about God's omniscience mean at the horizon of an open time? Well ... I believe it shows us that there are not only new aspects for humankind and in general for the universe but also for God. Does he know in advance about the almost incredible changes that chaotic processes could take in the course of millions of years? And can he discern from the beginning the events that result from fuzzy processes in the quantum world? How can one think about God in a time when the model of an eternal present, which also contains the past and the future, has to abdicate? Questions upon questions!

SHELDON: Take comfort, my friend. Remember that one of your former colleagues—Moses in the third of the Ten Commandments—warned about making an image of God.

But now we need to get to the control room. An error report from the unmanned moon explorer calls us rather rudely out of the enraptured heights of omniscience back into the saturnine reality. Out there, ODYSSEUS seems to have to tackle again something unforeseeable.

"It must have been rather unsettling when *Hermes Trismegistus* lost contact with the unmanned explorer ODYSSEUS from one second to the next," Randall said. "What happened?"

"We'll never know," said Hoihong. "I've studied the relevant data in the black box. The probe worked perfectly and produced fantastic pictures. Most likely it was hit by a meteoroid in the vicinity of the moon Phoebe. Meteoroids in the Saturn system originate through impacts by comets and asteroids on the moons. Fragments of ice are flung up and move on chaotic orbits until they hit a moon or, even more likely, Saturn's ring."

"It must have been frightening for the astronauts," Randall speculated. "They were equally exposed to such a danger."

But Astraia objected: "Astronauts are goal-oriented. Look at this postcard Sheldon wrote a few hours later."

The Beautiful World of Saturn's Moons
A Postcard from Saturn for the Cyber School

Hello, Earth!
Today I want to report on further results of the unmanned ODYSSEUS *probe. Before it fell silent, it explored the fantastic world of Saturn with a thoroughness never reached before. The majestic planet with about a hundred times Earth's mass is, as everybody knows, surrounded by more than eighty-two moons. Each of them is unique with its own story.*

The ice deserts shine brightly on Enceladus, *a moon three hundred miles in diameter. Its surface is covered with slowly flowing ice that originates from volcanoes where water and steam escape, freezing in flight into ice pellets and snow, and falling down onto the volcanic cone. Ice streams flow down the cone like glaciers, moving slowly away from the volcano. Wrinkled folds develop where the stream dams up. Some of the ice deserts are without any crater. They seem to be recent and must be less than one million years old.*

The ice streams on Enceladus have erased all traces of its earlier days. The climax of the flyby was a small eruption of a volcano known before. The column of liquid water soared vertically upward like a geyser and formed a gigantic umbrella of free-falling white ice crystals. It's not yet known where the energy of this spectacle comes from.

The somewhat larger ice moon Tethys *looks in comparison like an old man, although it's exactly the same age. With a diameter of over six hundred miles, it's one of the five large Saturn moons. Thickly covered with impact craters, it must still have, to a large extent, the original surface layer from its time of origin 4.6 billion years ago.*

The astronomers of the twentieth century have given Tethys's surface structures names from Homer's Odyssey. *Its largest crater is named Odysseus, like our probe. Melted snow filled the impact hole and is frozen into an ice plain. We scrutinized another remarkable geological formation, the Ithaca Chasma, with especial thoroughness. It's the name of a huge valley several miles deep, stretching some fifteen hundred miles over three-quarters of the moon's circumference. A gigantic impact must have happened in this moon's history, forming the Odysseus crater. The force of the impacting object probably almost blew the moon apart. Its interior warmed up and melted. The subsequent cooling and contraction left behind the deep rift.*

Dismal Phoebe, *the most distant of Saturn's large moons, is a peculiar misfit if only because of its orbit, circling counter to the other moons and the ring. For decades, astronomers presumed that Phoebe is a cuckoo in the nest of Saturn's family. Its surface is pitch-black, and if there hadn't been white places in the fissures and craters, it couldn't have been discovered from Earth.*

The black cover, however, is only a few inches thick and consists partly of organic molecules. They formed on tiny dust particles left behind by the vaporized water over billions of years and united with precipitating external dust into a tar-like soot layer. A shiver ran down my back when it became increasingly clear that Phoebe is a huge cometary nucleus.[2] With a diameter of 130 miles, it is about ten times as large as the object that hit Tethys.

Phoebe came perhaps from the depths of the universe. On its way to the sun, it was caught up in Saturn's gravity field. Luckily it got into an orbit without shattering any of its moons. If it hadn't been captured by Saturn, it would have continued its way into the inner solar system. The largest craters on Earth's moon, the Mare Orientale and the Mare Imbrium, resulted through impacts from comets or asteroids the size of Phoebe.

From Saturn with best regards, Sheldon Cutter

In the Shadow of Catastrophes

THOMAS: "Dismal" Phoebe didn't bring our good old ODYSSEUS any luck. No wonder the old witch turned out to be the nucleus of a comet. Comets are signs of coming disaster. The ancient people already knew that.

SHELDON: Yes, comets really can bring tragedy. Imagine a projectile of this size hits Earth. A crater measuring over two thousand miles in diameter would result, along with a tidal wave of several miles height washing around the whole Earth. The impact of the comet would have so much energy that the rocks would vaporize at the point of impact. The rock vapors then would explode and form a mushroom cloud hundreds of miles high. Dust would condense out of the vapor and spread out over the whole surface of Earth. Its heat would be enough to evaporate all the oceans so that the atmosphere's pressure inevitably increases two hundred times, with the temperature climbing over a thousand degrees. It would take thousands of years to rain out the water vapor. Earth would be uninhabitable for a very long time.

THOMAS: Your scenario takes my breath away. We can only hope this won't ever happen.

SHELDON: Ah, but it's already happened! In the first six hundred million years of its history, the young Earth was hit several times by objects the size of Phoebe. About fifty million years after its formation, even a veritable small planet the size of Mars scraped Earth. The dwarf planet was torn into pieces, and the debris mixed with the fractured material broken out of Earth's mantle, forming a disk around Earth. The gravity of the larger individual fragments absorbed the smaller ones and gradually everything massed together into one ball. The moon was formed![1]

THOMAS: Amazing! How awful and yet what a fine thing to have this beautiful moon as our companion.

SHELDON: An important impact happened sixty-six million years ago on the Mexican peninsula Yucatan. It not only extinguished the dinosaurs but also decimated three-quarters of all living species on Earth.

THOMAS: Yes, of course. Even I have read about that one. Thank goodness there have been no such asteroids since then.

SHELDON: No big asteroids, but Earth is still incessantly hit by cosmic objects, one hundred tons per day. But these objects are only seldom several yards in diameter. In Tunguska, Siberia, on June 30, 1908, it was probably a ten-yard fragment of a comet. No traces of a rock body were found that would indicate an asteroid. This particular celestial body was heated up during its flight through Earth's atmosphere, and it exploded while still in the air. The force of the explosion knocked all the trees around for fifty miles like match sticks.

THOMAS: Yes, but even though this explosion in Siberia was relatively modest compared to the earlier catastrophic events, it must have been a horrible calamity out of the blue for the local environment and perhaps also for some few Siberian hunters. Is a large impact on Earth to be feared in the future?

SHELDON: We don't know most of the cometary orbits exactly enough for precise predictions. In seventy-five years, on August 14, 2126, the comet Swift-Tuttle will come very close to Earth. The odds of it hitting Earth are only about one in ten thousand. But a solid impact by this celestial body— ten miles in size with a speed of around forty miles per second—would be catastrophic.

The near-Earth asteroid Apophis is another threat. It's named after an enemy of the ancient Egyptian sun god, because this villain will approach Earth many times in the near future. In 2004, the probability to strike Earth in the near future was calculated to 2.7 percent, but it missed Earth both in 2026 and 2036. Apophis is 270 meters in diameter; and its impact would release ten times the energy of Tsar Bomba, the biggest hydrogen bomb ever exploded. Apophis may well hit Earth some day in the future. But as Earth and the other planets interfere with its path, its motion is chaotic and therefore unpredictable in the long term.

THOMAS: So you believe that there's a distinct possibility of another catastrophe, and this time it won't be the vicious dinosaurs who are all destroyed but perhaps the cute squirrels or the innocent chickadees. The impact will cause dreadful widespread death. Who and what will survive? Will the insects finally rule Earth? Perhaps we humans have a good chance to survive, because then *the survival of the fittest* will apply.

SHELDON: Maybe not. With such huge and random death, one can't talk about a selection. Entire species will die out, including those that were previously so competent and dominating. In the great extinction at the end of the Permian geologic period, about 225 million years ago, when 95 percent of the marine species disappeared, the brachiopods, a group of sea animals that have hard shells on the upper and lower surfaces, lost their predominance in the sea to the mussels and never regained it. The catastrophe was probably caused by volcanic activity due to large continental drifts that led to enormous lava streams into the sea. The change of climate altered the flora and fauna so much that today, geologists can still use it to date their rock samples.

The dinosaurs "fitness" ensured them the supremacy for 130 million years on Earth, but it didn't help them sixty-six million years ago when an asteroid literally struck out of the blue. The small agile mammals made good use of the space cleared for them, while today the reptiles must be content with a marginal role.

THOMAS: With each large catastrophe the cards are reshuffled. It seems that less successful species are given a new chance in evolution. They step out of their shadowy existence with which they had to be content up to now. As you just said, the extinction of the dinosaurs triggered an evolutionary thrust with the mammals.

SHELDON: You still seem to be searching for a higher fairness with a secret agenda. I doubt there's a plan behind it. Whoever is affected by an unforeseeable catastrophe is pure chance. Often small details are decisive on how such an incident will have consequences on the other side of Earth, like the subsoil at the point of impact, the season, or the wind. A climatic catastrophe always follows large impacts because of dust-covered sky and heavy rainfall. The resources run short temporarily. Only now, "fitness" plays a decisive role because the rivalry among the survivors will be very high. This is when the most capable will survive. When the vegetation recovers, the conditions change again. Everyone can live well in the large free ecological niches, even unusual mutants.

In short, Darwin's evolution never plays out so rapidly as it does immediately after a major change. What then ultimately happens is, in the best sense, chaotic and not predictable.

THOMAS: Doesn't a basic principle of evolution show up here? A higher order establishes itself through a crisis or even a catastrophe, as unimaginable as it may have appeared before.

SHELDON: Yes, there are definitely long quiet phases with minor evolutionary dynamic, interrupted by massive developmental episodes. And they do often result from earthly or cosmic catastrophes. I presume, though, that you're still after a principle of progress. However, it's not even certain that evolution necessarily yields a higher development in a sense of an increase in complexity. There are numerous species of plants and animals that haven't changed for millions of years. Natural history is rather a large lottery; its symbol is the Dame Fortune's Wheel.[2] That must certainly be offensive for any religious understanding of the universe.

THOMAS: Not necessarily. Some religions consider the world as an inhospitable place, where sinister forces of fate or even blind chance lead the way. There are even actual catastrophe theories in some systems such as the classic gnosticism. According to them, the earthly world originated from a primeval accident. Contrary to this, Jews, Christians, and Muslims perceive nature as the work of a good Creator.

SHELDON: So here's the origin for your plea for higher development. It seems to come almost to an unholy alliance between classic Darwinists and Christian theologians.

THOMAS: No, I would feel as uncomfortable in the company of consequent Darwinists as with postmodern friends of evolutionary chance. Remember our Easter conversation. At that time, I was looking for an analogy between the storm in Saturn's atmosphere and the Easter event. I recognize in these events the crisis symbolized by the cross and the new creation in the resurrection, where a new order is established out of chaotic fluctuations. I think of evolution as a universal process that builds up new and more complex levels by striding through crises and catastrophes.

SHELDON: Chance plays a far more unpredictable role in the evolution of life than in star formation or in atmospheric processes where it is tamed by the law of large numbers. When many particles interact, the statistical average dominates and hinders large leaps. The randomness of gas dynamics in star formation and hurricanes originates in unstable initial conditions rather than the erratic course of events. In contrast to this, the impact of one single large asteroid can blow away all life forms that formed during millions or even billions of years on a planet. The paradigm "order out of chaos" incompletely describes evolutionary processes in biology.

THOMAS: I've never been so aware of how threatened and fragile the young life on our planet was for over hundreds of millions of years.

SHELDON: Evolution could actually have taken totally different courses. The influence of catastrophes can hardly be overestimated. I wonder whether you're so daring to find divine parables behind bombardments by comets and widespread death.

THOMAS: No, Sheldon, in view of catastrophes, past and present, it's more appropriate to become silent than to fuel meddlesome speculations. For me, the other side of Easter is equally important: Good Friday. With the cross, everything breaks off, and the resurrection symbolizes the absolutely not-to-be-expected miracle of a new creation. Isn't this a meaningful hint that God takes unpredictable risks with his creation?[3]

I don't want to remove the stumbling block that the cosmic catastrophes put in the way of a religious understanding of nature. Overcoming crises and catastrophes is an essential function in religion. We are in the midst of life surrounded by death, as it says in an old church hymn by Martin Luther.[4] Creation is no ornamental garden. Nonetheless, it can become our home in which we feel safe and secure.

SHELDON: You remind me about speculations whether comets brought the germ of life to Earth in early times.[5] But now I notice that the Titan crew has found and repaired the defect in the life-support system. Not every breakdown means a catastrophe! ORPHEUS's landing time, however, is postponed for another several hours due to the glitch.

THOMAS: How will they choose a safe landing site?

SHELDON: A yard-thick layer of polymer sediment covers wide areas of the Titan's surface. But they've just chosen a landing site in a young crater that the Titan crew spontaneously baptized "Crater of Good Hope." Presumably it was created by the impact of a comet fragment. The energy of the fallen celestial body melted the ice of Titan's rock. A lake of several miles filled the crater, froze again, and made a smooth sludge-free landing location. The target is in the northern hemisphere, not far from the Kraken Mare, the largest sea on Titan's surface.

THOMAS: For me, the concurrence of the landing on the giant moon with the loss of the ODYSSEUS probe stirred up dark forebodings in my dreams. I'll post it in my dream diary, so it can become part of our official record.

Figure 9. On January 14, 2005, the *Huygens* probe landed on Titan and took this picture. The two rocklike objects just below the middle of the image are about 6 inches (left) and 1.5 inches (center) across at a distance of 33 inches from *Huygens*. Some objects are rounded like pebbles eroded by a river. They are suggested to be frozen water ice. (Photo: ESA/NASA/JPL/University of Arizona)

Groping in the Depth
From Thomas's Diary (May 13)

Tonight I dreamed that I was abducted into Titan's methane fumes. I floated slowly through the defuse clouds and fogbanks. The sunlight, still blazing brightly despite its distance, became weaker and weaker. The fog thickened to titanic cloud formations. Deeper and deeper, I dove into a strange reddish shimmering world. Sheet lightning pulsated in cloud formations.

On the edge of my field of vision, outlines of a dark gigantic city with high-reaching towers, monstrous temples and palaces, cyclopean walls, and yawning gates were flitting about. As I looked on startled, deep-orange fog wavered as if it wanted to play a nerve-racking game with me. I was afraid as I advanced into this unfathomable world, where strange beings lived in titanic cities and worshipped nameless gods. The deeper the methane vapors drove me against the eons-old brooding secrets of Titan, the more I could feel their weird dreams. I woke up distraught.

Landing on Titan
From Sheldon's Diary (May 14)

The ORPHEUS *has landed! I can hardly believe that we've finally attained the goal we worked toward for so many years. I'm writing in my diary for the first time in months, so thrilling and significant is this accomplishment that I've been privileged to witness from my perch in the* Hermes Trismegistus.

Five people in bulky protective suits move on Titan's strange dark-red landscape. We experience it up here like seafarers in the sixteenth century who had to remain onboard the mother ship and joyfully, anxiously, and yearningly watch their colleagues in the landing boat as they set foot on a foreign continent for the first time.

I absorbed every piece of information eagerly, no matter how trivial and long known. The measurements suddenly begin to live. Now I can feel the coldness of -290°F, sense the low gravity of one-seventh the value on

Earth, experience the low visibility and the lack of mobility. Although I can't really contribute anything, I shared the thrill of getting the electro-sledge going in the icy wind. I was overcome with a deep feeling of happiness with the description of the bizarre ice rocks at the rim of the crater. We've made it!

At this moment I feel at one with the magnificent adventure. We are no longer individuals who live and suffer through their individual fate. No, a historical event is happening! Humanity on Earth, even the whole solar system, is also part of it. I'm not only a spectator on the grandstand, but I'm taking part in what's happening and react with great joy and, at times, also fear. What I hear about the Titan expedition touches me deeply, although my own safety is not affected.

My strong reaction to the landing must also have something to do with the fact that it's a central event in my life. Not like an action item in my daily agenda but my target for ten years—what I lived and worked for. With the successful landing, my commitment receives meaning.

I have never felt such happiness about an objective research result, however important. So I'm surprised all the more about the effect of this landing. There was no scientific discovery, no new data, no new theory. But it has had a psychological result that far surpasses objective measurements. It's a different kind of perception when I take active part as a subject.

It's hard to believe how my view of the world has been changed today after the three years of interplanetary travel. I seriously wonder whether the universe in the long run was made for the joy of beings like me.

Is happiness the goal of the cosmic development? What nonsense! Evolution is still one horrible bloody drama. How can both be valid, death and life?

Do such participatory perceptions supply answers to questions like this? Can one learn to handle them?

Perhaps the good Thomas Haubensak tries to answer these questions with his theology, but his language remains strange to me.

What a day!

Extraterrestrial Intelligences and Their Religion

THOMAS: Listen, Sheldon. The landing of ORPHEUS has provoked a major response from religious and theological sources—both individuals and institutions—throughout the world. The question of possible extraterrestrial intelligences was also discussed on Earth by renowned religious panels and organizations. While official Protestant statements were marked by strong restraint, if not quite reluctance, the Roman Catholic Church emphasized its reach to *omnis creatura*, all creatures. The Eastern Orthodox Church also indicated a lively interest. No reactions were issued from Islam, while the Hindus and Buddhists called to mind their own universalistic traditions.

SHELDON: Yes, Thomas, and three years ago, the media's special attention focused especially on the sixteenth Dalai Lama, who not only resorted to the old theory of Buddha's manifold manifestations in the different universes but also certified Buddhism a special affinity to a cosmos-wide religious awareness.

THOMAS: The Dalai Lama has always appealed to me. But do you remember the Irish tabloid that called me the "Missionary of the Calvinistic God"? Now our colleagues have landed on the mysterious Titan. How I would have liked to be there! I love this cocktail of fear and joy of discovery more than everything else.

SHELDON: Well, we'll both land on the ice moon Iapetus in less than four weeks. From there, Saturn offers us an overwhelming view. No veil of haze will cover up the shining stars. Do you appreciate that no other priest has ever come so close to God?

THOMAS: You really ought to know better than to tease me. We're not one step closer to God in the outer region of the solar system than on our home planet, Earth.

SHELDON: Then perhaps you're a missionary on the Saturn mission?

THOMAS: I could live with that label if *mission* isn't meant as a more or less violent conversion of members of other religions but rather a testimony shaped by the figure of Jesus that leads to setting up a conversation between human cultures.

SHELDON: Testimony of what? Of God?

THOMAS: A testimony of the experience that our life represents a precious gift, with something new streaming toward us every day. A testimony that expresses something like an invitation to engage with this life stream without reserve.

SHELDON: I don't feel very comfortable when our evening talk mutates into a missionary situation. Of course, you vigorously shake your head, but take a look through our porthole! Myriads of celestial bodies, stars, even galaxies and world systems! There might be numerous planets accommodating intelligent beings. Perhaps tomorrow our friends will get in touch with intelligent Titan inhabitants. Can you imagine conducting an interreligious dialogue with them?

THOMAS: There seem to be religious experiences in all human cultures. If it's about perceptions of a depth dimension of reality, we could talk about them with any aliens.

SHELDON: I'm not so sure that extraterrestrial civilizations will be a comfortable area to sow for earthly religions. How would you make the biblical narrative about Jesus accessible for them?

THOMAS: We'd have to translate the Bible's concept of God into their language and in their understanding of reality. The old vision of a Christ who fills the whole cosmos could be helpful for bridging that kind of gap.

SHELDON: What does that mean? It reminds me of our Easter discussion, where you even wanted to look for the Son of God in a storm on Saturn.

THOMAS: It's exactly that perspective that attracts me. The belief in the cosmic dimensions of the divine Christ was very decisive in early Christianity. The Son of God may reveal himself in many ways, according to the third-century theologian Origen of Alexandria: to the humans as a human, to the angels as an angel. With this view, the incarnation of Jesus would be a specifically designed revelation figure of his diverse being for us earthly people. Origen talked about the "accommodation," the adaption of the divine manifestation for the respective receiver.[1]

SHELDON: Are you suggesting that the human Jesus is only a subset of the total cosmic Christ?

THOMAS: Set theory was no longer *en vogue* in my school mathematics, but your notion makes sense to me. One can't talk anymore about one incarnation. According to Origen, Christ also acquired the figure of an angel. Unfortunately, the old church rejected his theory as heresy.

SHELDON: That's typical! Always this intolerance among the religious!

THOMAS: I also regret the condemnation, although I much appreciate the concern of the church fathers to protect the unique figure of the earthly Jesus.

SHELDON: So you'd permit still other incarnations for the aliens.

THOMAS: Yes, Sheldon. It's conceivable and even fascinating for me. Imagine how God would reveal himself to intelligent giant jellyfishes in the warm oceans of a planet of brilliant Aldebaran.

SHELDON: As a jellyfish?!

THOMAS: Why not? Or he might manifest himself to the silicon brains in the vastness of the Andromeda galaxy as a pulsating silicon crystal.

SHELDON: Your power of imagination may have blown a fuse. Let's put aside the jellyfish of Aldebaran and Andromeda's crystals. A subtle inconsistency in your formulations strikes me. First you talked about the cosmic Jesus. Now it's God who reveals himself in numerous figures. Do we have a set theory here again? Is Christ a subset of God?

THOMAS: The fine shift of terms inadvertently breaks open subtle issues. When I speak about a revelation of Christ, even if he's also a cosmic divine dimension, my focus is narrower and sharper than when talking about a revelation of God.

SHELDON: Certainly! Many religions talk about God, and even some of my colleagues, especially those advanced in years, can identify God behind cosmic laws or fine-tuning. On the other hand, the name of Christ refers specifically to a Christian concept of God.

THOMAS: Precisely! Here is a sharper and, therefore, narrower focusing.

SHELDON: And what does this clearer focusing tell us? What would be the specific characteristic of the subset "Christ"?

THOMAS: God's focusing onto Jesus Christ boils down to comprehending God's nature as *love*. With Jesus, God comes to us locally and in the limits of space and time. A large number of people who led miserable lives are touched by the unconditional compassion of God in encountering Jesus. His ordeal up to the horrible death on the cross testifies that God's love makes inroads in all abysses of forlornness.

SHELDON: The plurality of religions reminds me of a marketplace with many vendors, where each tries to drown out the other. So there are many different theories about the early universe that compete with one another. Everyone who works on a theory is convinced that it's the correct one. Only a few will be directly disproven. The less successful simply are gradually forgotten and finally smiled at by later generations.

THOMAS: The difference between religion and science is important to me. For me, "truth" in the religious sense can't consist of a system of timeless valid principles, but is rather an experience as one proceeds along the way in life.

SHELDON: But there are many issues where religious programs are mutually exclusive! Talking about religious interpretations in the analogy to scientific theories, it's conceivable that one will become unsustainable or revised due to new insights.

THOMAS: Dispute is part of the dialogue between religions. In such a situation, for example, I must give reasons why I prefer the Easter paradigm over other interpretative patterns for the deciphering of the world.

SHELDON: So, if our colleagues who just landed on Titan meet intelligent inhabitants, surely there will be exciting discussions. Whoever lives so close to Saturn perhaps has quite different interpretations to offer for its atmospheric structures.

THOMAS: I hope so! Then we'll see whether Titanian theology understands the creation processes on Saturn more deeply. If this is the case, I would like to take a continuing education course on Titan. It would well fit the impending Pentecost, the feast of the spirit, to which I'm looking forward in any case.

SHELDON: We'll soon find out, since our fellow astronauts are exploring Titan with the CHARON caterpillar, the Carrier for Humans and Retrieval of Objects of Newness. Look at these live pictures from the bizarre crater rim they are sending from "Good Hope." After an extensive circle around the Titan lander, astronauts Takeo and Shadia have ventured up the steep rock formations of the nearby southern rim of the crater to set up a camera. The atmospheric conditions are excellent compared with the usual circumstances on Titan, and the crew expressed their enthusiasm that the light-red clouds repeatedly broke up and produced fascinating light shows in front of a violet background.

THOMAS: I'm inspired to record my emotions about these stunning images in my diary.

Pentecost Is Near
From Thomas's Diary (May 14)

ORPHEUS's *successful landing on Titan stimulated an exuberant euphoria in us and then a deep relaxation. May our colleagues return unharmed into our cosmic sphere lit up by the radiance of the distant sun after the exploration of this mysterious underworld!*

Orpheus, the great singer in Greek mythology, succeeded on his voyage to Hades to bewitch the cold hearts of the dark rulers of the dead with his charming lyre music. They allowed him to take his lovely wife Eurydice, who died far too early, back into the daylight. Our friends in the modern ORPHEUS *craft have no singing muse at their side; their instruments are microscopes, mobile laboratories, and data processors. They don't bank on the seductive power of magic singing but on the ability of experiments and analysis—not on empathy and participation but on distance and objectivity. Perhaps they'll also bring evidence back into our human daytime world of a wondrously strange life down in the methane-clouded depth.*

The undertaking of the ancient Orpheus, however, failed. Overcome by doubts and longing, but contrary to the laws of the rulers of the realm of the dead, he looked around for his loved one too early and she was snapped back into irrevocable death. May a similar adversity not befall the modern ORPHEUS *team! Cool matter-of-factness will save them from becoming confused by ruinous sentiments like the old singer.*

With our voyage to Saturn's system, we've ventured farther into space than anyone before us. Now we enter the veiled world of Titan that is incomparably more mysterious than foreign islands and distant continents in antiquity. It's a place where dreams and visions haunt me. Only the monotonous procedure of our regulated daily routine and the discussions with Sheldon committed to rationality controls and stabilizes the turbulences of my agitated imagination.

In one week, we will celebrate Pentecost. It strikes me as a happy coincidence that this feast of the Holy Spirit is happening just as we've reached the peak of our expedition. I'm reminded of the old prophetical words of Joel—of promises that already began to be fulfilled at the feast of Pentecost with the first Christians:

> Then afterward
> I will pour out my spirit on all flesh;
> your sons and your daughters shall prophesy,
> your old men shall dream dreams,
> and your young men shall see visions.
> Even on the male and female slaves,
> in those days, I will pour out my spirit.[2]

Israel's prophet expected the marvelous outpouring of the Holy Spirit in the midst of huge cosmic catastrophes, from "wonders in the heavens and on earth, blood and fire and billows of smoke."[3] The sun turns to darkness and the moon to blood, until God himself appears in all his might and splendor.

For the early Christians, the cosmic turning point took a different shape. The outpouring of the Holy Spirit at Pentecost was a kind of "big bang." The gospel of Jesus Christ begins to spread out to all people around the world, up to the most distant islands and coasts. The Spirit kindles the hearts of the people for the message of love of the Creator that embraces everything in new spaces and times.

The thoughts Sheldon and I exchange in our evening talks stimulate me to tell stories. My own spirit swings far up and away beyond the borders of the solar system as if I felt the breath of life-creating divine spirit in the wings of my soul.

Shimmering jellyfishes endowed with reason pass in front of my drunken eyes in their majestic circles in a warm ocean under the fiery-red light of the huge Aldebaran, a red giant among the stars. At the rim of our Milky Way, a single plant has spread out over the whole surface of a planet—a collective superorganism, existing of countless but totally integrated intelligentsia. It has already begun to spread out its spores into the universe in view of its dying sun, protected by sophisticated *membranes from the deadly cosmic radiation, to arise one day to new life in another world.*

Intelligent life on the basis of silicon crystals builds up in the distant Andromeda galaxy before my mind's eye. Millions of years will pass for it as fast as days do for us. These creatures literally live at the beginning of the end-time. They will experience firsthand whether the expansion of the universe comes to an end or accelerates even more. Finally, my fantasy creates rational entities that have totally said goodbye to the coarse-matter transient world of crude subatomic hadrons. Immortal photons are their carrier medium, and their bodies are made out of pure light . . .

Here, where our sun has been reduced to a little glowing spot and our blue planet can barely be seen with the naked eye, I am once again aware that with the dawn of the modern age, the dimensions of space and time have broadened. The exorbitant size of today's observable space pushes our good old Earth into an indifferent area of the universe—orbiting around a commonplace star in the nowhere of an average galaxy. Despite the objection of the evangelist Luke, the history of Israel and Jesus indeed took place "in the farthest corner" of the world, at a totally different level than the people of old could have imagined in their wildest dreams.[4]

The people of the Christian late antiquity and Middle Ages assumed they lived only a few thousand years after the creation of the world. In their eyes, the Christ event was easily something like the middle of the time, in a strange coincidence concurrent with the first decades of the Roman imperial rule and the peace in the Mediterranean area brought by Augustus. For us, the abysmal depth of time has opened, and the history of humankind represents only a short flare-up in the passage of immense eons.[5]

The biblical history of salvation that reaches from Abraham to Jesus to the time of the early church threatens to evaporate out here into the cosmic vastness. I wonder more and more: Can the Deity really have bound himself so exclusively and definitively to our Earth—at a certain time in the Quaternary period, and to a specific place, near the rift valley between the African and the Arabian continental plates?

On the other hand, I can't agree with a pure cosmic religiosity that totally disregards history. It makes more sense to do without an absolute position of "objective" time and space and to take the perspective character of religious statements seriously. The historical contentions from Abraham to Christ, in which the biblical God reveals himself, would then have for us *an authoritative revelatory reality. For* others, *God would make himself known in a totally different shape, in the field of their own experiences with the cosmos and history. These different perspectives indicate, each in its own way, an interconnecting and all-pervading divine One.*

I notice my heart beat and the movement of my breathing. How natural—and how wonderful! It's like a perpetual prayer taking place deep within me; the famous everlasting Prayer of the Heart from Mount Athos in Greece. Doesn't every breath call for the divine breath; doesn't every heartbeat bear witness of the great life? Am I not constantly invited to participate in these movements with vigilance and devotion to make their unconscious prayer to a conscious praise of God?[6]

"All things pray except the First."

The uplifting expression of the Platonists goes through my mind. The whole cosmos, permeated with all-encompassing sympathy, takes part in one uninterrupted prayer. Just as everything streams from the One, so it returns back to its origin. This return is nothing but a hymn and a prayer, accomplished by every form of life. If we listen to the universe with the ear of the spirit and look with the spiritual eye, we become aware of this all-fulfilling song of praise. It sounds in a mighty choir—from spiritual beings to humans, to plants and to stones.

For the old Platonists, the cosmos is permeated by numerous series of analogies. The sun's series stretches from the highest sun of the spiritual cosmos down to the rooster that greets the star every morning, to the lotus and the sunflower. Similar analogies are in the signs of other gods, like Cronus-Saturn. Its series reaches from the angelic beings of the highest wisdom and cognition, over to the hermits and melancholics, down to the ibex, roots, bones, and lead.

I'm fascinated by the Platonists' teaching of all-cosmic analogies and parables. The highest deity projects himself in always new manifestations on each level of the cosmos, in an uninterrupted series of archetypes and images.

For antiquity and in the Middle Ages, it was God who wrote the images and parables of himself into his creation. Not only theologians and poets but also naturalists were engaged in deciphering this book of creation and to find God's image in it. The circumstances reversed with the change into modern times, when it became man who invents images and parables. The tie between image and being is torn; only the human imagination remains to bridge provisionally the irrevocably separated.

For many modern observers of nature, this "mystic" vision of a cosmos singing praise is shattered. The first to notice was Blaise Pascal in the seventeenth century, when he wrote that "the eternal silence of these infinite spaces frightens me."[7]

I find it hard to share this depressive view of a largely dead universe. Should organic life be a strange unwelcome guest in a deadly cold universe? Doesn't the universe rather press for life in whatever colorful forms and configurations it may take? Doesn't the divine light refract even in each atom and in each elementary particle?

Life—the Most Beautiful Child of the Universe

SHELDON: Thomas, I have some exciting news from the astronauts who landed on Titan.

THOMAS: What is it?

SHELDON: They've discovered a steaming warm spring on the eastern rim of the crater in the ethane snow at the foot of an ice rock. Pablo measured a water temperature of 40°F at ambient conditions of 288°F below zero. Amazing what a medical doctor can do! The source seems to be of volcanic origin. Microscopic surveys of the highly methane-containing colorless liquid revealed microdrops—that is, droplike structures that dissolve only at a temperature above 70°F.

THOMAS: I'm surprised that such exciting news reached us so shortly after landing. The formation of life out of inanimate matter is, for me, the epitome of an act of creation. Indeed, a work of the creator God that can be grasped with hands.

SHELDON: OK, Thomas, but let's stick to the facts. Such microscopic droplets were an important preliminary stage of life on Earth. Their membrane regulates the exchange with the environment. Within that protective surface, biochemical reactions take place in a controlled and uniform manner.

THOMAS: In the midst of a cold world, hostile to life, a wondrous order emerged. Do you know the beautiful old Christmas song about the rose that begins to bloom in the middle of winter?[1]

SHELDON: Yes, but the time of year doesn't fit. Besides, life doesn't develop specifically like a rose. Only a blind watchmaker[2] is at work, and he doesn't systematically combine piece by piece into a whole. Most of the droplets are nonviable and wrongly constructed. They soon pass away in Titan's icy atmosphere and dissolve in the disorder of the surrounding molecules.

THOMAS: I'm still surprised how a totally new order could arise. Before, the molecules moved around independently in the atmosphere and the slush, and now they build a fascinating structure. These droplets open up a new dimension of reality. Something qualitatively new is formed.

SHELDON: No, it isn't, my friend. These droplets are practically the same molecules. If the suitable conditions are given, the new order is inevitable. Living cells belong to the most complex phenomena of the universe known to us. Nevertheless, they proceed according to the same laws as the formation of a storm on Saturn.

THOMAS: So conditions alone can produce living organisms?

SHELDON: No. Actually, new entities develop all the time. They arise by themselves. Like in a hurricane, there is no need for an external stimulus that causes the development. If a small disturbance in Saturn's atmosphere forms an upward movement, it may get stronger without outside influence. If there should be a God at work here, then his activity is restricted to playing dice. The possibilities are limited. Chance makes the decision.

THOMAS: I can't get used to the dice-playing God. So back to the inner coherence of the world: I'm impressed how evident the analogy becomes between such different kinds of phenomena such as a hurricane and a prebiotic molecular process.

SHELDON: More than a hazy analogy exists here. It's the same mathematics. Everything can be ultimately traced back to the elementary processes in the quantum world. It's true that small systems behave periodically in the quantum mechanical description. Some theorists have concluded from it that, contrary to our world, the quantum world isn't chaotic. Far from it! In large quantum systems, the motion of the virtual and real particles is overlaid with chaotic dynamics.[3] The microenvironment's chaos is much closer to the macroscopic reality than one could imagine. It's the virulent interior of the macrocosm.

THOMAS: That's amazing! Our firm and spacious world is permeated by the ecstatic dance of the quantum world and its fluctuations. In our processes of consciousness, we have a direct part in this basal chaos. Didn't a famous physicist even derive human freedom directly from the indetermination of the quantum world?

SHELDON: Pascual Jordan, a German mathematical physicist who made significant contributions to quantum mechanics and field theory back in the mid-twentieth century, suggested it.[4] I remember very well a talk show ten years ago where a science theorist heaped mordant scorn on this rash marriage between physics and philosophy. There's no way the strict laws of probability amplitudes of quantum theory lead to the subjective sense of freedom.

THOMAS: I don't want to enter the steep terrain of philosophy of freedom either. Yet I still wonder if fantasy, intuition, and perhaps even mystic conditions originate by participating in the infinite ocean of the quantum world.

SHELDON: You let yourself be seduced by the suggestive power of images and metaphors that we physicists use for illustrating the abstract world of our mathematical formulas. What strikes me, however, is a turnaround in your thinking. First you slandered the search of the physicists for a coherent theory as reductionist. Now you grasp for that kind of derivation. Is this only a matter of taste? Evidently you find dancing quanta more congenial than stiff atoms or iron laws.

THOMAS: Indeed, the paradigm of cosmic spontaneity and unpredictability appeals to me more than a dead unreeling world machine.

SHELDON: I feel the same! It's fantastic to see an open future before us. I've experienced it more than ever before at the landing of ORPHEUS on Titan. This has nothing at all to do with religion.

THOMAS: Each of us participates and has his own reaction, Sheldon, as I'm sure you'll agree.

SHELDON: I do. And here's a new communication from the Titan crew. They've set up a research station with an electric generator, oxygen, and water supplies on the eastern rim of the Crater of Good Hope. A permanent video connection to *Hermes Trismegistus* was installed.

THOMAS: Can we call the station *Elysium*? It was the name of the place where the blessed go after death in classical mythology, a kind of mysterious paradise.

SHELDON: Well yes, we can. But it seems like just another of your efforts to draw metaphors and analogies between physical truths—such as our fellow astronauts setting up a scientific research station on Titan—and the supernatural.

THOMAS: Physics and theology sit cozily here in the same boat. They're both heirs of the ancient Greek metaphysics. Unlike many of my colleagues, I never got involved with the trend toward ideological pluralism. And I would never dump and bury the old conception of unity of the world. The theoretical physicists of the last hundred years indulged in that speculation no less than the early philosophers. Today they are metaphysicians par excellence.

SHELDON: I don't agree at all with that. Our construction of a theory isn't simply wildly growing uncontrolled speculation. The longing for world unity, though, connects us indeed.

THOMAS: Perhaps the enormous distance from our native Earth, our being thrown out into the endlessness of the universe, pressures us into an unholy alliance. We are both children of Saturn, the star of the speculative philosophers. Here's another entry in my diary that was inspired by the pictures sent by the research station.

A Vision of Titan
From Thomas's Diary (May 15)

Once more, with great pleasure, I download the pictures of the crater walls transmitted from ORPHEUS *onto my cabin monitor. The bluish-green structures rising into the dark-red sky are finely chiseled and remind me of the external bracings on Gothic cathedrals. The ethane ice has built indescribably bizarre sculptures.*

I envy how my co-astronauts will dare to pass by these elaborately wrought crater walls at a suitable place in the next days. How magnificent it must be to finally arrive at the shore of the methane ocean, to watch the gentle surf and the sheet lightning above the crimson surface with their gigantic icebergs. Will they also see there on the horizon the violet glistening mountains out of pure ice, higher than all summits on Earth? With shivers, I remember the famously weird science fiction novella At the Mountains of Madness *by H. P. Lovecraft.*

These pictures accompanied me while I fell asleep. There! I look at a surging ocean, with colossal icebergs jutting out of it. At their bottom they continuously melt and freeze again. The more I let my view wander up to the peak of the icebergs, one rising above the other, the more diverse and artistic the crystalline structures become. As I float closer to the pinnacles and towers of the crystal palace born out of the sea and touch the cold grandeur, my finger doesn't encounter ice but senses flowing water, a mighty bubbling waterfall whose spray even sprinkles my face. Just as I barely pull back a little, the waterfall freezes again into crystalline pinnacles.

I awaken abruptly and stretch out in my hammock. What was that? Had my mind allowed itself to be stimulated into a natural philosophical allegory? Did it want to take shelter in elementary and vivid symbols from the abstract concept formations of last evening's talk?

The interpretation of the dream seems easy. Of course, there is the ocean of timelessness with its fluctuating quanta, a ceaseless play of creation and dissolution of all matter. The cosmos grows like a tower out of this flood, an unbelievably complex structure made of time and space that keeps piling up higher and higher. More and more, the past amasses, more choices are made, more structures are frozen. And yet! Beneath its apparently unchanging surface, there surging and streaming, a play of pure energy, free of all heaviness.

And more—this constant dance at the base of all being inflates, faster and faster, the whole universe with enormous power. What does it mean if there's an elementary connection between the growth of space and the openness of time? The perpetual undulation of the quantum ocean expands the dimensions of our space-time, in which our unique history takes place.

"And God Saw That It Was Good"? Creation Is Evaluated

The ORPHEUS crew carried out further chemical and prebiotic analyses the day after the great discovery of the microdrops. A constant stream of information flowed between the Titanauts in the caterpillar to those in the landed spacecraft. The *Hermes Trismegistus* received pictures and listened in: chemical composition, temperatures, forms and sizes of the drops, further macromolecules, and much more. The daily routine gradually returned again and with it a day-after mood.

"I don't understand why Sheldon and Thomas, both illiterate in biology, got into the following vivid discussion about life," Randall asked his colleagues.

Investigative journalist Astraia knew the answer.

"I searched in the sciencetainment server and found that the two Hermenauts downloaded the multimedia documentation "Greatness and Atrocity of Nature" from the Department of Biological Sciences at the University of Montreal. The film seems to have stimulated them to an extended conversation."

THOMAS: My goodness, this film was really a bit much! I'm afraid it will keep me from sleeping tonight. Nature has spawned so many unbelievably absurd, monstrous, and brutal beasts that one could go mad.

SHELDON: I feel exactly the same way. I knew about these things theoretically. But to be confronted so directly and sensually with it, exposes my nervous system to a brilliant firework of neurotransmitters.

THOMAS: With this cue, you just reminded me of that drug dealer in the insect kingdom, the beautiful large blue butterfly, that I know from my childhood vacations in the heathland of Swabian Jura. The butterfly attracts whole troops of ants with a sweet secretion. The ants then drag their own brood to the greedy large blue caterpillars that feed the tender meat of ant larvae to their own spawn. In their intoxication, the ants don't realize that these caterpillars devour their offspring.

SHELDON: How do you bring all these phenomena, that you name cruel or grotesque, together with your belief in a benevolent Creator?

THOMAS: That's a difficult question. The kindness of the creator God and his world rule are also drawn into doubt by humanity's atrocities and horrific actions. Here, one could blame the errant human freedom, deviating from the divine will. In nature, the problem of theodicy becomes even more unavoidable. Why did God create so much evil? That's why the Gnostics in antiquity claimed that not the good God but an inferior angel created the world.

SHELDON: This angel today has a new name. It's called evolution.

THOMAS: I adhere to the biblical creed of God as creator. There is nothing in the world where he isn't present. But his countenance is sometimes hidden.

SHELDON: We already debated this question in our Easter conversation. While you mask your aporia with the metaphor of the "hidden face," the evolutionary paradigm offers an understandable solution for all the horror in nature. What this film so ruthlessly presents is quite expedient in evolution and derives from the interplay of mutation and selection, from chance and adaption. As repulsive as many things in biology may be for us, all are explained by their evolutionary origins.

THOMAS: Perhaps we have to better understand the mechanism of biological evolution in order to reconcile how evil can exist in a world created in divine goodness. But you remind me that besides our human viewpoint and that of nature, there's still a third to be considered: God's perspective. Also, in the field of religion, one can often make the mistake of equating one's own perspective with God's. The priestly account of the biblical creation narrative in which God's goodness is praised seems at first sight to take a distinctly anthropocentric perspective. The divine works are therefore "good" because they benefit humankind.

SHELDON: Biology derived from Darwin's approach is in the position to explain by simple mechanisms many astonishing phenomena in the world of living creatures. You would have to assume that God also uses these ground rules.

THOMAS: Some creation traditions proceed from repeated experimenting of the gods. According to the Mayas, only the fourth trial was successful when creating humans. Creation of the man's companion didn't succeed right off, either, according to the Bible's older creation narrative; animals appear as the first, still unsatisfactory "help" for the man. When later the wickedness of humankind grew outrageous, God was "grieved" that he had made humankind and he brought the Flood over the earth.[1] In this old stock of ideas, it seems there is also apparently trial and error on the part of God. Later, these elements were totally left out when the concept of God's perfection was elaborated.

SHELDON: That reminds me of your reanimation of old chaos myths when we discussed about the vacuum and the divine creation of new things. You could even picture God as playing dice. But all of this culminates in the dismissal of God's omnipotence.

THOMAS: We need to clarify what we mean by *omnipotence*. I won't get lost now in the history of Middle Age theology or in a philosophical analysis of the word. You're right in so far as the inherent dynamics of the world, impressively emphasized by the natural sciences, affects God's image. God no longer seems to be at the helm of cosmic evolution. That challenges numerous religious convictions.

SHELDON: In your view, God creates and keeps the universe in its entirety from the outside, so to speak. Shouldn't he have any possibility to also influence things in the inside of the world, in the details? Does God preserve the universe like a ship from sinking in the sea? Can't he have any control on its course, because everything runs autonomously inside of the ship? Is this truly your opinion?

THOMAS: Here I can only grope. Jews and Christians sometimes thought about a "self-emptying of God," something like his "self-renunciation."[2] In ancient times, the rabbis spoke about God's self-humiliation, following his people into exile and serving them like a slave. Christians discussed the incarnation of the Son of God, arguing whether Christ totally gave up his divine nature in order to join us as a limited weak human. Occasionally theologians even speculated about whether God divested himself of his omnipotence at the very beginning, to give the world the possibility for its own existence and growth. What a tremendous idea—God limits himself in order to grant his creation its own space and time.

SHELDON: That implies the autonomy of the world. There couldn't be a divine intervention anymore. You've moved far away from traditional religious statements and the Bible.

THOMAS: The Bible talks about both: God's wonderful or subtle controlling *and* that things in history have happened that God didn't really want. Though, it is true. The idea about God's "self-emptying" breaks with many basic religious convictions. It deals with a shimmering and also dangerous teaching. It not only tempts us to underestimate God but also to overestimate humans. Theologians are well advised to handle such dynamite with great care.

SHELDON: My thought is whether one wouldn't be better off to give up the hypothesis of "God" totally. If he lingers "outside" without being able to influence creation, then not only science but also religion can do without him. A God who doesn't hear or answer prayers and doesn't promise any rescue miracles isn't very admirable or appealing. He reminds me of the gods of the Greek philosopher Epicurus. My grandfather raved about Epicurus's philosophy, where the gods didn't interfere in the world but led a carefree life in the realms between the universes.

THOMAS: Epicurus was more religious than one generally thinks. But you're right that a god who totally departs from the world can neither be experienced nor worshipped. The notion of God's "self-emptying" aims at exactly the opposite. God's abandonment of his unlimited rule over the universe doesn't mean his absence in the world but his increased presence.

The way of his presence is described as walking with us, living with us, suffering with us. This is no longer a God who stopped the sun and moon for almost a day during a battle to support his people. However, here is a God who walks with the pilgrim and goes before him like a shepherd: "Even though I walk through the darkest valley, I fear no evil; for you are with me; your rod and your staff—they comfort me." And when Jesus in the darkest hour of his crucifixion utters the prayer "My God, my God, why have you forsaken me?" he's assuming that God can hear him.[3]

SHELDON: Saturn's sphere seems to have moved you into the darker aspects of life. Your image of God seems appropriate for wailing old women.

THOMAS: No, not at all. I not only sense God's presence in sorrow and pain but also joy and happiness.

SHELDON: Saturn's cold world still rules over us, and for a long time to come. I feel torn between this planet and Earth, between the desire to explore the universe and the longing for my feet on the ground. But back to our topic. We can't determine the footprints of God anywhere, neither in the cosmic background radiation nor in evolution's episodes of life on Earth. That's why you want to deny God's influence in the universe. With that, you reduce him to a role of an invisible ghostly companion who is present everywhere but who actually participates nowhere in the world game.

THOMAS: God takes the people into his presence. He develops faith, hope, and love in us. He makes himself known to the souls with intuitions, inspirations, and revelations. Remember our Easter discussion when we talked about participation, visions, and ecstatic conditions!

SHELDON: So now you're thinking very dualistically—God has no effect on matter, but definitely on the "soul." I fear you're jumping out of the frying pan into the fire. I don't know what the "soul" is. I prefer to talk about a complex neuronal system that can be examined by exact methods.[4] Not only instincts and feelings but also visions and intuitions are based on neurochemical processes. If God is able to produce certain endorphins in the limbic system, he can just as well manipulate the genetic information of the earliest archaea.

THOMAS: You fall back again to the most obvious reductionism. Neuroscience doesn't see the difference between a subtle artistic experience and a mass euphoria during a boxing match. As a student, I loved to look into the old alchemy, and I hope to forge some interesting links at this point.

SHELDON: Alchemy? Instead of an alchemist turning lead into gold and similar charlatans, I expect much more from the promising breakthroughs in modern quantum neurology. After all, we're in the middle of the twenty-first century.

God—Mighty King or Poor Wanderer?

SHELDON: I find it hard to bring your two main points into relation with each other. On the one hand, the creator God incessantly sustains the universe; on the other hand, he's supposed to have substantially limited his absolute power.

THOMAS: That is indeed the crucial point. Both statements apply necessarily and simultaneously to the concept of God . They are the poles of an indissoluble field of tension. There is the aspect of the immense might and majesty of God: he creates and preserves his creation at every new moment. And there is the opposite aspect: he limits himself and gives creation space and time for its own course. Here, he is the magnificent king, where the worlds run into nothing before him and decay like dust if he doesn't consistently breathe in his creational spirit, as we read in Psalm 104:

When you hide your face, they are dismayed;
when you take away their breath, they die and return to their dust.
When you send forth your spirit, they are created;
and you renew the face of the ground.

But there he is the unprotected wanderer who walks along with his willful creation wherever its path may go, as Jesus promised at the end of the Gospel of Matthew: "I am with you always, to the end of the age."[1]

Omnipotence and powerlessness are closely related with each other in many biblical texts.[2]

SHELDON: I see no connection, but rather only a contradiction.

THOMAS: I understand. It's a seemingly paradoxical thought that the mighty king and the poor wanderer are one and the same. Theology has developed complex reflections regarding the self-differentiation of God that ultimately all revolve around the Trinity doctrine. Out here in Saturn's sphere, vivid metaphors are closer to me than theoretical operations.

The king and the wanderer: What unifies both sides of the Deity? I believe it is God's outpouring love. The king has the loving heart of the wanderer. His creative force originates from pure love. He maintains the continuation of the world process with endless effort in this love. And out of the same love he grants his creation wide space and endures having to relinquish its paths to the fickle play of blind chance and the constraints of iron laws.

SHELDON: I'll gladly forgive you the images instead of sharp dialectic. We physicists also resort to illustrations to make the formalism of our theory understandable to a wide audience. But my main impression still remains that you make God smaller. You limit his omniscience and omnipotence. The uncertainty of the quantum world and the coincidences of evolution press hard on the old gentleman.

THOMAS: Here I have to contradict you. God's self-limiting for the sake of his creation gives enormous latitude to time and space. He bound himself to the skills of his creation in his love. Space in the form of unpredictable quantum fluctuations and time in the form of nonlinear processes make the creation project a risky adventure. Yes, God has absolutely become "smaller" here. But that's only one side. The flipside is the monstrosity of dimensions that our modern perspective unveils about the universe. Space and time have positively exploded before our eyes. With them, the God who creates and maintains all this grows into the immeasurable. Again, I recall the depth of the cosmic chasms, the scale of the staggering proportions that open up by the ascent through the hierarchy of the cosmic structures.

SHELDON: You mean the mental jump from the solar system to the Milky Way, from there to the Local Group of galaxies, further to Virgo's supercluster, right up to the even larger galaxy formations that structure the universe like walls of a honeycomb. All of this is only the observable part of the universe. Possibly it's still substantially larger than the segment that is accessible for us, not to mention other hypothetical universes.

THOMAS: Yes! And God is again greater and deeper! Mystic texts from ancient Judaism envisioned God's grandeur by a seer ascending through the different hierarchies of angels up to the divine throne. In each case, every higher class outdoes the preceding one beyond all measures in splendor of light and holiness; the seer falls from unconsciousness to unconsciousness. Nevertheless, all these hosts of angels are only ephemeral whiffs before the divine throne. God's transcendence is thereby enhanced with this into the unimaginable. Today's recognizable extent of space has a similar psychotropic effect on me.

SHELDON: Especially since you can also extend the scale downward, down to the smallest. At the birth of the universe, all of these unbelievable spaces were compressed into the order of magnitude of subatomic particles.

THOMAS: Precisely. It's similar for me with the time dimensions. Here, too, every humanly imaginable measure is surpassed. The difference of the time spans that were comprehensible for people of antiquity and the Middle Ages compared to our view can't be overestimated.

SHELDON: At least the endlessness of space was already postulated before the beginning of the modern era.

THOMAS: Yes, theology tried to take this up in physicotheology of the seventeenth and eighteenth century. The pure hypothetical assumption of an infinite space with perhaps innumerable world isles is very different compared to today's quite precise knowledge of hierarchically nested structures and the timescales of the universe. In the twentieth century, the cosmos has not only become quantitatively a trillion times larger but has also gained a totally different shape. We haven't really explored what that means yet for the experience of creation. I suspect it implies a great magnification of God, contrary to this "scaling down" that we just talked about. How great God's creational workmanship has become in view of all the cosmic chasms and abysses of space and time!

SHELDON: Really? Amid a remote galaxy, why should our small planet be of significance at all for God? Not to mention individual believers with his or her wavering daily problems.

THOMAS: Because this great God became "for our sakes" low, small, and helpless. He is a wanderer who walks with us. Seen this way, "omnipotence" has nothing to do with the arbitrariness of a potentate. As we say in our jargon, it has its place in the *doxology*, in praising God. That is, in the usage of prayer, singing hymns, and thanksgiving. It is the perspective of people who come to experience themselves as those receiving a gift without any reason.

SHELDON: Yet, in the theological speech about creation, God is usually praised as the big player.

THOMAS: But the experiences of creation go together with thanks and rejoicing, fear and worries. In Psalms, statements of almightiness answer experiences of protection and deliverance:

> *The Lord is king, he is robed in majesty;*
> *the Lord is robed, he is girded with strength.*
> *He has established the world; it shall never be moved.*[3]

The worshipper here, who awards God's kingdom and power, has his place within creation. He doesn't relocate himself from the world of experience out into a distanced, quasi-godlike position from where he could measure out God's capacity to act. What you abstractly postulate as "omnipotence" is a chimera. It has nothing to do with the experienced God on whose exertion of power a prayer or hymn answers.

SHELDON: Once again you seem to give feelings and emotions a far higher importance than my objective point of view. Evidently this is also your position for the understanding of God's omnipotence.

THOMAS: Some of my colleagues are in the habit of speaking of the almightiness of *love* that also includes powerlessness. I even play with a bolder possibility of understanding. Omnipotence reflects something that in the present tangible world does not yet exist. In antiquity, Jews and Christians were convinced that God's kingdom would only be all-embracing and fully completed in the future. Our glimpses of God's power, celebrated in praise and thanks, would be a preview only of that which one day will come abundantly in completion.

But it might be better to completely avoid the misleading term *omnipotence*. I consider not only the prefix *omni* in need of explanation, but I even resent applying to God the ambivalent term of *power*.

SHELDON: Did you say "completion"? I pricked up my ears when I heard that word. We should discuss the future of the universe.

The Universe's Hospitable Nature

THOMAS: Apparently the universe pushes inexorably toward life. While surfing in our video library during the long flight to Saturn's system, however, I was astonished to learn how many special conditions have to be fulfilled. For me, this unique design of the cosmos is an important indication for a divine intelligence that arranged all this.

SHELDON: Slow down, Thomas. I fear you're running in circles, trying to catch your shadow. When it became clear in the last century that the universe developed from a dense and hot initial original state in the big bang, astrophysicists wondered how the dynamics of a developing universe could lead to the formation of life. Of course, very special conditions are needed, but they were obviously available. Otherwise we wouldn't be here and able to ask why.

THOMAS: Yes, you know better than I how special these cosmological circumstances are. The correct type of star, the right kind of planet at the right distance, composed of rocks that contain the necessary minerals, with a lot of water, adding a moon to cause high and low tides, and many more things were needed so that life, as we know it, could be formed. The development of life is not at all a matter of course, let alone the rise of intelligence.

SHELDON: We're getting into an argument regarding probability again. Don't forget that there are more than one hundred billion stars in our galaxy, the Milky Way, alone. In the whole universe, there are at least one hundred billion times more galaxies. There are more planets than stars, and probably over ten thousand billion billion planets in the universe. The necessary conditions could be several times fulfilled with so many possibilities.

THOMAS: You refer again to a strange coincidence. Elementary constants are the same at all times in the entire universe. There's no selection of possibilities here. Chance can't play at all. With only a single shot, the chances would certainly be extremely small to exactly meet the correct values.

SHELDON: Perhaps physicists in the future can deduce and explain these constants with an overriding theory of why precisely these values have made possible the development of life. In any case, there is no reason to suspect the orderly hand of a God behind this coincidence. I believe another hypothesis is more appealing. Maybe there are many universes, in which the elementary constants have different randomly distributed values. Then we wouldn't need to wonder that we live in a universe that's suitable for the biological development. Otherwise we wouldn't be floating here.

THOMAS: You're offering me two unattractive alternatives. The first one consoles me with possible future results in physics, for which there are neither prospects nor indications. The second explanation seems to me even more dubious. How can one put the universe, the epitome for everything and the sum of all parts, into multiplicity? Doesn't the term *universe* include all possible or at least all observable objects? All other hypothetical universes would not be observable as a matter of principle. Since when is it allowed again in physics to introduce unobservable quantities?

SHELDON: The only observable fact is the fine-tuning in initial values and constants. It is an indication for a higher quantity and the selection of universes.

THOMAS: I understand that the accepted methods of physics don't allow the hypothesis of God. With the shadow universes, however, you introduce unobservable entities that cannot be objects of science.

SHELDON: Fine-tuning as proof of God's existence is not conclusive. For instance, the homogeneous distribution of galaxies at large scale was an enigma for a long time. How can the density in a region of the universe so far away from us have the same value as in the opposite region, where the light of both only reaches us now in the middle of the distance? Both parts of space can't exchange information because of the finite speed of light.

This uniform distribution was considered to be an initial condition for the cosmic development, requiring fine-tuning. The proposed theory of a superluminal inflation phase in the early universe in 1980 offers a natural and relatively simple answer. The expansion of space was faster than the speed of light for a short time. Before that inflation, space was much smaller and interactions adjusted the density in all parts of the universe we can observe today. Consequently, the alleged proof for the existence of God went up in smoke.

THOMAS: I don't want to capitalize on gaps and peculiarities of scientific facts. Nonetheless, these apparently improbable cosmic parameters for life and intelligence make me think. Don't we run into a puzzling boundary of our knowledge? This is not a *proof* for the existence of God, but very likely a significant hint. In this light, even one's own life seems less self-evident, but finely tuned.

SHELDON: Cosmic fine-tuning doesn't necessarily indicate a plan. Physicists discuss the hypothesis of a *goal-oriented tendency* without ulterior theological motives.[1] This tendency would be a quality besides the established symmetries and laws. It could be similar to entropy, whose increase also exhibits a goal-oriented character: the second principle of thermodynamics states that the entropy of a closed system can only stay the same or increase, whatever the causes are. The same would be true of the tendency toward the emergence of life. Yet, causal or random reasons must still exist.

THOMAS: I note that science and the belief in a creator meet without searching for causal explanations. The nexus is the amazement about the expediency of the universe. Your hypothesis of a general cosmic goal can be expanded into a comprehensive view through the religious interpretation.[2]

SHELDON: I don't believe that the universe is heading inevitably for the goal to develop life. The final tendency seems to be too good to be true. It's all the more dangerous to force theological interpretations upon questionable scientific hypotheses. What a shaky construction! How can you make it plausible for someone like me?

THOMAS: I don't start from physical models but from existential experiences. The course of my life is composed of an innumerable succession of coincidences. Many could have been different. All the same, when I look back, I can't avoid the impression of some providence at work. My life's journey in all its twists and turns was accompanied—in religious terms, guided—by a deeper sense. Guided even to this place in the mysterious world of Saturn. All this gives me hope for the future. In light of these experiences, the world appears to me as gift. I'm touched by the secret of creation.

SHELDON: The IASA's search committee for the mission's astronaut candidates would be very irritated to hear that they, of all people, should be regarded as an instrument of divine guidance. But seriously, sometimes I also discover my life is finely tuned. Despite all the isolation, I'm thankful to participate in this space expedition. I interpret this phenomenon differently—namely, as a favorable strategy of survival. Those people who learned to get used to the facts were more successful and preferred by evolution.

THOMAS: You're a hopeless Darwinist, Sheldon, from top to toe!

SHELDON: Then you are a fatalist, Thomas. Every little happenstance is the will of God.

THOMAS: Yes, but a fatalist, as you imagine him, has the same attitude toward the future as the past. That would be a truly fatal error. When I look forward, there is no plan but a wide space for an open future. When I look back, I detect a meaningful order that couldn't have gone any other way. We can't go beyond this sharp difference of time modes. Therefore, talking about a plan of God, whether it's on an individual or a universal level, is totally wrong. It would deteriorate into the delusion to know this plan or even be able to execute it, as if a human could place him- or herself in God's position.

SHELDON: I can't discover a plan of God in the universe either. Evolution isn't working toward a goal, but blindly gropes its way forward. Nature really does work like a blind watchmaker.[3] That's why I'm skeptical about the hypothesis of a cosmic goal. It seems to me like a last theological relic in the guise of physical language.

THOMAS: The last diffuse fog, soon to be resolved by the rising sun of Darwinist super theories! I've already provoked you once with my suspicion that your inclination for total theories still lives off an inheritance of metaphysical speculations. Long ago, theologians bid farewell to the arrogance of knowing God's plan. Today, systems theorists and theoretical physicists have taken over this position.

SHELDON: There are, in fact, some physicists who lay claim to the inheritance of theology.[4] Yet the overwhelming majority of scientists keep to the observable. No indications can be identified there that evolution is heading for a distant goal. Teleology, the doctrine of goal-directedness and functionality of nature, has abdicated for good.

THOMAS: We come here to a point where modern science and theology probably confront each other in a basic antithesis. As a theologian, I maintain that creation not only comes from God but also goes back toward him. Formulated more pointedly, God even comes toward it. It's not enough that God called creation into existence and from then on accompanies it. He approaches it.

SHELDON: Are you talking about the flow of time from the future into the past?

THOMAS: Exactly. This coming of God can't be manifested on the level of causality, which produces the future from the past.

SHELDON: Then you should speak more precisely of a lack of contact instead of an antithesis between science and theology. For me, this higher level definitely remains ghostly—the classic example of an unnecessary hypothesis. Does this other dimension even appear on the level of natural phenomena?

THOMAS: Despite your skepticism, I stand with the old teleology and would like to vary my image. God comes toward his creation by pulling it to him. He's like a central star whose gravitation draws everything to it. His "drawing" from the future may happen in different ways. The intertwined pathways on which the single parts of the world arrive at the Deity aren't fixed and may run a bit "chaotic."

SHELDON: Are you thinking of the "prime mover" of Aristotle?!

THOMAS: Absolutely. But the God I'm talking about is full of compassion for his beings, who let themselves be pulled by him. His force of attraction is love. He draws his people like the father does in the parable of the lost son. The son lets himself be moved by thoughts about his good father and looks for the way back to him. As a result, he has a wondrous experience. The father rushes toward him full of compassion and gives him all that he has.[5]

SHELDON: You appeal to me with your peculiar interpretation of that old story more than the TV preachers who place the emphasis in this parable totally on remorseful repentance and active penance. For my part, I would have more readily thought about Odysseus, who searched his way back home on intertwined, winding tracks.

But hang on. We're getting some communication from the ORPHEUS crew on Titan.

THOMAS: What's going on? Everything OK?

SHELDON: They've had to battle the ethane hoarfrost buildup on the radio antenna. But I'm able to continue to monitor and report the observation programs requested by ground control from the Arrayed Reflector Gadget for Optical Studies.

THOMAS: Do you mean the one-hundred-foot telescope ARGOS on the mother ship *Hermes Trismegistus*?

SHELDON: The very same. I can direct the telescope toward Earth . . . our home . . . only half a hand width angular distance away from the sun. And look beyond Saturn's system as well.

THOMAS: I don't blame you for the nostalgia for our planet Earth or the curiosity about what's out there beyond Saturn. But I remember them telling us that an inadvertent pan over the glaring sun would subject the equipment to considerable risk, like melting the sensitive detectors. Besides, remember that we weren't sent to Saturn to amuse ourselves with hobby observations.

SHELDON: You're right. I confess to indulging my nostalgia and curiosity, but I am going to write a postcard about it.

Homesick for Paradise
A Postcard from Saturn for the Cyber School

Hello, Earth!
I can't conceal any longer that I recently had a look at Earth, the blue jewel in the deep-black sky, through the giant eye of the Hermes Trismegistus. *These days, Earth is in its "greatest brilliancy," thus far away from the sun and well illuminated. Our one-hundred-foot telescope* ARGOS *can not only recognize continents, oceans, and mountain ranges but even larger cities.*

When suddenly the veil of clouds disappeared over my home area of North Dakota, pictures of my youth came to mind as we romped across blooming meadows and dove into cool rivers. How simple life was then! Everything necessary was there, right at hand. There was no need for bulky spacesuits or complicated breathing systems. Much is taken for granted on Earth that causes daily worries and troubles here. Despite our immense knowledge, no engineer could build such a perfect spacecraft as Earth.

Seen from our location, Earth is only a half a hand width away from the sun. We're here at nine times the distance from the sun to Earth. The outmost rim of the solar system, the edge of the Oort cloud of comets, is ten thousand times farther. I tried to look out there for gatherings of such snow bodies. Hopeless! At this distance and in this darkness, even ARGOS *is totally blind.*

Darkness and icy coldness hostile to life govern in the tremendous spaces of the outer solar system. The hot solar wind doesn't make inroads into these regions. The interstellar gas already catches it a thousandth of the way. The sun's radiation is so weak out there that it can no longer noticeably warm up a celestial body above the temperature of the universe, −455°F. The next star, Proxima Centauri, is only three times farther away from the outer edge of Oort's cloud.

The abundant possibilities for life on Earth are wonderful gifts of nature. Do you know, dear people of Earth, on what a unique marvelous masterpiece you dwell in the midst of the desolate vastness of the universe?

From Saturn with best regards, Sheldon Cutter

Götterdämmerung of Humanity?

THOMAS: The pictures from our large telescope really touched me. Our Earth is a perfect dream.

SHELDON: We don't need any diving suits to move about outdoors on Earth. The blue planet is truly a paradise in the huge universe. I've rarely felt it so intensely as in this icy world of death in which we've gone.

THOMAS: But your pictures also show the relentless ever-increasing devastation: gigantic forests burned, widely polluted tracts of land, carbon-laded air and toxic bodies of water. We humans have become a cancerous ulcer in the living organism of the biosphere.

SHELDON: Yes, Thomas, the colossal distance to our home planet shows us that humankind is only a product and a part of evolution. When an animal species has stripped its habitat bare and destroyed its basis of existence, its population automatically decreases. It will finally die out or adjust itself. I'm hoping that humanity solves the problem before that catastrophe becomes reality.

THOMAS: I share your hope and fear. Before humanity is threatened with extinction, numerous plants and animals will be obliterated. More than a hundred of the thirty million species die out every day. If that continues, we'll be all alone in a few hundred years . . . or sooner.

SHELDON: I pray that humans will survive today's phase of extinction by further developing themselves.

THOMAS: I see few opportunities at the moment for a biological evolution of humans. In the past ten thousand years, our kind barely changed physically.

SHELDON: Perhaps humans will further develop their intelligence. The most intelligent is also the most capable in a technical civilization.

THOMAS: That depends on how one defines intelligence. Whether we humans really act intelligently in the long term is very doubtful. Our technical supercivilization has become a threat for the whole biosphere. I fear we're stumbling into a self-made catastrophe.

SHELDON: I'm more optimistic than you. There were far bigger upheavals in the past. About two billion years ago, a gigantic catastrophe descended on the world of microorganisms when Earth's atmosphere became enriched with oxygen. Yet evolution continued. The chemistry of the organisms gradually adjusted itself. Today, our successful Saturn mission shows the enormous innovative capacities of our technology. I trust completely our practical rationality. In the human consciousness, the universe thinks perhaps for the first time about its own future. Such nonlinear feedback allows new dimensions of development. Evolution has always regulated itself; humankind can also do so.

THOMAS: I'm not clear on what you base your evolutionary optimism.

SHELDON: The situation of the past million years is a lot like the time of the middle Cambrian period 510 million years ago, when evolution discovered the advantage of multicellular organisms. In the world of single-celled eukaryotes, a new dimension opened and new living beings developed. According to a prevalent theory, all of the approximately one hundred important animal phyla formed in only a few million years. In the meantime, seventy of those phyla have already died out, but no new ones were added. Since then, the multicellular organisms with their complexity and decisive biological advantages have carried evolution further. Similarly, evolution discovered a new dimension and direction, inventing, for instance, the human consciousness.

THOMAS: I sense a diffuse religious remnant in your trust in evolution. I would be more pessimistic here from the natural history perspective. Humans originated in the shadow of global catastrophes, and their end may be like that as well.

SHELDON: So-called catastrophes in some cases caused higher stress levels with positive results. The cooling off and the downsizing of East African woods decisively contributed to the development of the *Australopithecus* some six million years ago, as well as further climate changes later to the forming of *Homo habilis* and *Homo erectus*. The intensified competition among the human species at the time of the Pleistocene may have provoked the enormous development of the cerebrum.

THOMAS: I shudder when I think about the emergence of these rapacious hordes of hunters of the Ice Age. Humans were hunter-gatherers for over two million years. Only a wafer-thin varnish of a short civilization phase lies over this archaic dark legacy of hominids. Occasionally it quickly gains the upper hand today.

SHELDON: You overestimate the violence potential of hunting societies. The opposite is true. Indications of violent conflicts become more frequent only with the development of agriculture ten thousand years ago.[1]

THOMAS: That reminds me of the first murder in biblical prehistory—the quarrel between the nomad Cain and the farmer Abel. Violence and war increased dramatically with the forming of advanced civilizations. Yet the threat to all animal species by the early humans is obvious. There are many alarming hints, for example, that they are primarily responsible for the extinction of the large mammals on the American continent, where there once were mastodons, giant sloths, and large armadillos.

SHELDON: You're thinking of the Pleistocene overkill. In fact, many large animals at that time suddenly died out, not only in America. Humankind also exterminated numerous other great animal species before the beginning of modern history, such as the giant birds in New Zealand. This destructive potential has increased massively with the technical revolution. But I'm not sure what you're driving at. Are you looking for symptoms of original sin? Are you totally pessimistic regarding the future of humankind?

THOMAS: I don't want to make myself an advocate of original sin. It is an ambiguous construct. However, I am disturbed by the dark shadow that followed humankind from the beginning. Everything cumulates into the seemingly unsolvable ecological problems that modern society almost inevitably produces.

SHELDON: You can talk! You're sitting comfortably in the most ingenious and sophisticated machine ever built by humankind, and you defame modern technology.

THOMAS: This insolvable conflict is exactly what bothers me. The possibilities of our science and technology fascinate me no less than they do you. The staggering price frightens me all the more for what we have to pay for these delicate fruits of the tree of knowledge. Despite all my passion for space travel, I can't agree with your claim that our *Hermes Trismegistus* is the height of our accomplishments. You've overlooked the grand potential that human cultural development provides. It offers us complex software for the supercomputer in our head—for our brain. Culture gives us self-awareness and uplifts our consciousness.

SHELDON: I'll gladly agree to that. The question is, can we reprogram the circuits in our brain before it's too late? I'm hoping for a rapid evolution of human intelligence.

THOMAS: That feels like the ghost of eugenics. I doubt that a perfection of the cognitive intelligence will help us further. Just the opposite! Our main problem is that our rational thinking separates itself more and more from our older intelligence. I can only hope for a true development in the cultural domain, where it comes to individual integration between younger and older parts of the brain, and where the rational functions are harmonized with those of emotions and feelings. The human consciousness may evolve if the historically older part, which is still very dominant, can be integrated with the younger evolutionary achievements. I think of something like a better exchange between the limbic system, the location of emotions, and the cortex, the seat of our specific human consciousness and thinking. To provoke you a little more, I believe that this integration is cultivated predominantly in the realm of religion.

SHELDON: Please explain.

THOMAS: We can probably get through the present-day challenges on a long-term basis if environmentally acceptable behavior isn't only dictated by our rational considerations but shared also by deeper impulses. Insight and feelings make space for what many old texts call "wisdom," integrating oneself in the orders and dimension of the earthly environment instead of breaking it up. Here, religion could supply an invaluable service.[2] Albert Schweitzer spoke about the "reverence for life." He meant a kind of religious attitude, based on the ethics of solidarity with the other living creatures.[3]

SHELDON: So what's the relation to religion?

THOMAS: The awareness of belonging to a single global community has very much to do with religion—religion understood as a culture of dealing with the gift of life. In the ecological context, it means to respect other types of life and to grieve over today's creeping disappearance of so many colorful life forms. Religious ethics tends to oppose the evolutionary survival of the fittest.[4] Not the strong but the weak can be selected and promoted. It doesn't rely on merciless selection but on empathy, compassion, and solidarity.

SHELDON: I doubt whether religion can make people more sensitive for ecological connections. First, it would have to take notice of the scientific facts. I agree, however, that ecological maxims shouldn't address only our cognitive capacities. This insight must be considered in children's education. Chaos theory gives me hope: small causes can have large results.

THOMAS: That's an encouraging closing remark. My limbic system is now vehemently calling for some sleep. But before I get into my sleeping bag, I'm going to write in my diary.

Titan's Shadow over Humankind
From Thomas's Diary (May 17–18)

The pictures of our Earth that we can get from such a great distance affect me more than I could tell Sheldon. Is our biosphere on the way to becoming a single huge city? A name literally forces itself into my mind: Titanopolis!

Did we have to distance ourselves so far from our home planet into the realm of the old Titan gods to find the real name for the earthly project of modern people?[5] *In the last hundred years, humanity actually made itself equal with these archaic gods when it determinedly left behind the boundaries that until recently were thought to be set immovably. Revolutions shook up the four worlds, as the philosophical tradition used to divide them up from time immemorial: the worlds of matter, life, soul, and mind.*

We had acquired power over the plutonic forces of elementary particles, over the creative plans of biological construction, over the physiology of our central nervous system; and we finally achieved the ability to create our own electronic world. The ancient promise of the snake—"and you will be like God"—has been fulfilled in our midst.[6] *I'm also exposed to the seductive power of Titanic might. We have willfully passed a distinct borderline in our mission to Saturn's system, and paradoxically we are thus moved into the sphere of the ancient gods, of these professional border crossers.*

From Babel to Titanopolis

But my fascination for technology gives way more and more to shivering and horror. In the shadow of hubris lurks the fall. There was the fateful ship with the highly symbolic name Titanic *that steamed into its own doom and should excite the fantasy of the people. Long before that, though, there was the gigantic Tower of Babel, when in the ancient Orient the first advanced civilizations developed with their claim of encompassing world dominion, as I read in Genesis:*

*Come, let us build ourselves a city, and a tower with its top in the heavens!*⁷

Jewish legends say that the Titans or giants of ages ago built the city in order to attack heaven from the tower. But I read in Proverbs,

*Pride goes before destruction, and a haughty spirit before a fall.*⁸

Ancient people still had the instinctive knowledge that there is a sound measure and a sacrosanct limit for everything. Our species has totally stripped off this dread. Will the fall that follows the attack on heaven be dreadful? When will the call ring out: "Fallen, fallen is Titanopolis the great!" ⁹ *The violation of the face of our wonderful Earth and the brutalization of the people show very clearly what is brewing. Will we find our way back in time to a culture of moderation? Can our titanic passions and our sharp intellect—both products of our archaic history as dangerous predatory apes—be recast into the figure of a real human being, a person who is truly created in the image of God?*

A Dark Variant of the Creation of Humankind

*Recollections of dark mythological narratives regarding the creation of humankind pass in my mind's eye—about Babylon's god Kingu, huge and butchered; and the overthrown rebellious Titans and giants, from whose blood humans were made. The Orphics' ancient dark myth of humankind's creation, sung by Orpheus, also does not leave my mind.*¹⁰ *He recounted that we humans carry a dark element in us since our origin: Zeus killed the abominable Titans with thunderbolts; and from the soot he formed humans.*¹¹ *The sinister power of the Titans extends into our innermost parts. Now, around midnight, the most famous of all descendants of the Titanic race comes into my view. Prometheus, the son of Iapetus and Themis, who is said in an alternative creation story to have made humans out of clay and is regarded as their cunning patron.*

At the dawn of human history, he secretly stole fire from the palace of heaven for his protégés and therefore incurred the horrible penalty of the god-king Zeus. He was hung on a mountain, forged at the end of the known world for an almost endless time, torn to pieces daily by an eagle. Heracles is said to have finally freed him.

What a memorable myth! The greatness and the tremendousness of humanity are contained in this figure of the philanthropic Prometheus, whose name means "the one who knows in advance."[12] *His fire created the path for us humans to rule over Earth more than one million years ago. And now, as I write, we could never have penetrated into Saturn's empire without the fiery ion thrusters of our spaceship, his gift to our ancestors.*

Plutonium energy cells preserve us out here from the deadly ice world. However, in his gift, the Titan fire of nuclear power, slumbers also a destructive force. Alas, the "before-knower" could not convey to us the power of thinking that could have made possible wiser dealing with his gift, far ahead in time and space. Thus he brought over us blessing and curse. For eons he is chained on the merciless hard matter, at the end of the world, with open wounds, crucified for our good.[13] *Does this suffering benefactor announce the figure of Christ from a distance?*

Zeus feared Prometheus not only for stealing the fire from Olympus but because Prometheus knew about the end of Zeus's austere patriarchy in the future thanks to his mother Themis. Prometheus alone knew the future divine child who would wrest away the kingship from Zeus, like he himself violently robbed it from his father, Cronus-Saturn.[14]

I grope for the meaning of this old myth. Can the sequences of plateaus and crises in the evolution of human culture be interpreted with the antique theory of eons? The strict and inexorable rationality rules in Zeus's era, which conquered the history of humankind with the Greeks. The formerly ruling mythical powers, represented by the Titan king Cronus, were banned to the depths of our unconscious. Yet the beginning of this Olympian era was a nefarious venture: Zeus gained the power in the same way as his father, Cronus, once did. The ancient inheritance of the Titans still threatens at the foundation of the immaterialized and purified rule of Zeus. The suffering of Prometheus reminds me that even in Zeus, the god of intellect, the archaic violence of the usurper is still alive.

The Titan Prometheus represents not only the long past but the still eerily lively. As a "before-knower," he looks into the future and beyond Zeus's rule. While the rational mind represented by Zeus cannot see its own limits and claims eternity, Prometheus knows, thanks to the wisdom from his mother, what is hidden in the bosom of the future.

What is it that he sees coming? Do the irrational forces of the Titans return again? Or does the shadow of its dark inheritance no longer befall the new epoch of the world? Will even a wondrous powerful goddess rule Earth in kindness and compassion, and free life forever from the curse of the Titanic violence?

Sheldon had dreams in those days that dig deep into his emotional and mental states. They reflect the difficulties of the mission to Saturn. Here is one that he showed Thomas, the resident psychotherapist, for analysis.

At the Dawn of the Last Days—an Apocalyptic Dream
From Sheldon's Diary (May 19)

Last night I had a horrible dream that follows me today like a pack of yapping dogs. The sun was in a cloudless sky above blooming meadows on beautiful Earth. Suddenly I had the impression that the sun was becoming larger and redder. First I thought it was only an atmospheric phenomenon, as at sunset. Then I became hotter.

The sun grew and grew. It absorbed all the water from the trees and even from my lips. Out of the lakes rose steam; the flowers withered. The sun had the size of a fist and grew larger. Meadows and forests started to burn. There was no more water for extinguishing. The asphalt on the streets formed a gluey mass where cars stuck like flies on an adhesive strip.

People couldn't flee, and they put on protection suits made of insulation material. They reminded me in their appearance of our colleagues on Titan. A glowing hot wind blew from the sun. It was now larger than a stretched-out hand, red and mighty like Titan through our porthole, and it inflated more and more.

The heat became unbearable. The oceans evaporated and distributed a gruesome dampness all over Earth. The sight became cloudy. Patches of steam darkened the sun, only to be seen as a reddish spot in the sky. Earth became a dead desert, immersed in dull red light. It was a scene as on Titan. The air pressure also rose noticeably and left me breathless. The people panted as if under a heavy burden and their voices sounded like those from ORPHEUS. *The heat forced the people to retreat from Earth's surface; they could only survive in deep caverns. The houses burned down and the concrete melted like wax in the sun. Factories, libraries, and churches caved in. Whole cities collapsed and became a viscous mass. In the caverns, there was no electricity or food.*

Finally, the heat became too high for life in the underground shelters as well. The sun grew further and soon covered a quarter of the sky. And it grew faster and faster. I had the impression of plunging into it in free fall. I woke up wet from sweat.

Figure 10. Radar image of a landscape near the eastern shoreline of Kraken Mare, a hydrocarbon sea on Titan's surface. The image was assembled from observations by the *Cassini* spacecraft in 2007. It spans 120 miles, and the tallest mountain is about 3,600 feet high. (NASA/JPL-Caltech/ASI)

In the Chat Room of the Journalists (June 19)

"At this point, the information from the dialogues becomes meager. One more day to go until the deadline for submitting our report," Randall sighed. "The last three days of the mission to Saturn known to us are most relevant. They begin with the premonitions of Sheldon's dream."

"No problem," interrupted Astraia Callas. "Since the events on Titan were recorded in minute detail by *Hermes Trismegistus*, its electronic archives give us relatively detailed material about the following thirty-six hours. Let's focus on the recordings of the official log, except for purely technical data."

"And always bear in mind, Randall," Hoihong Wong cautioned, "the extremely hostile cold environment for the exploration of Titan's mysterious "ice desert."[1]

"Yes, yes," Astraia Callas added from London, "but the landing was soon after sunrise, so the crew had almost eight Earth days of uninterrupted sunshine at its disposal, half a rotation period of Titan."

"Thanks to you both," Randall said. "Here's the summary that I prepared last night from the transcripts of Sheldon and Thomas's technical dialogues."

After a long delay, the landing of the five Titanauts took place in the Crater of Good Hope (CGH), a relatively flat young comet crater about four miles in diameter. It borders a small volcano already discovered from orbit in the south. The volcano is possibly connected with the comet's impact, and it seems to have been active a short time ago. Perhaps the cosmic impact has made a connection to the hot subterranean water layer inside the moon.

According to some new maps, the Kraken Mare intrudes inland to the neighborhood of the landing site. This bay in the northeast was named Sinus Numinosus.

After the Titanauts anchored the landing craft ORPHEUS against the wind gusts and made it ready again for takeoff, they explored the crater and its surroundings with the caterpillar CHARON. On May 15, they made the sensational discovery of primary stages of life at the eastern crater rim, where a mighty mountain range of partly melted rugged rocks of water ice arises. They set up the permanent station *Elysium* at the warm methane spring and visited it regularly during the following four days. At the same time, they set up a methane energy plant for testing purposes next to the landing craft.

On May 18, Shadia and Nicole, the computer scientist and the biochemist, climbed the central mountain in the north to which they conferred the Titan name Mount Atlas. Thanks to Titan's sevenfold lower gravity, both hopped enthusiastically in wide jumps to the observation point that revealed a fascinating view of the entire crater. There they installed a fully automatic relay station for radio communication as well as a further monitoring station with a camera.

In the meantime, after a long troublesome search, the other crew members found a passage in the northeastern crater wall through the ice rocks. During the remaining two days available before the departure shortly before Titan's nightfall, a three-member group wanted to reach the coastline of the ocean.

The time was already considerably advanced due to the unexpected intensive work in *Elysium*. Also, a change of weather was on its way. So far, they had quite good meteorological conditions for Titan circumstances. Yet the instruments on Mount Atlas, as well as the observations of the troposphere by the Doppler meter and infrared bolometer on HTM, led them to expect increasing stormy conditions with high precipitation.

Sheldon's suggestion to cancel the expedition to the sea was rejected by the Titanauts since a further delay of this long-awaited operation would have made it impossible in view of the night's fall in three days. Although he wasn't responsible for this decision, Sheldon was troubled by feelings of guilt.

"Certain things are simply not allowed on Titan," he reflected later thoughtfully, referring to Orpheus's failed underworld journey, which Thomas often recalled.

"And here are the logbook entries," Hoihong said, "of the two days after that (time is given in UT)."

Logbook Entries from May 19–20

13:17 A team of three—Sergei, Nicole, and Takeo—set out with CHARON on the expedition. Target area is the coast of the large methane ocean northeast of the crater, the Sinus Numinosus bay. The distance from ORPHEUS to the passage northeast of the CGH crater rim is 3 miles, from there to the ocean 8 miles. The two remaining astronauts, Shadia and Pablo, return with a transport sledge to *Elysium* to bring 624 pounds (weight on Earth) of valuable soil samples to the spaceship and begin the first preparations for the takeoff scheduled for May 20. Both teams have contact with each other as well as with the crew on *Hermes Trismegistus* via the relay station on the central mountain.

13:55 Increasing clouds drift into the area of CGH; methane precipitation sets in.

14:23 Sheldon Cutter from *Hermes Trismegistus* suggests the CHARON team abort the expedition because of bad weather forecasts. Sergei decides to continue.

14:47 CHARON reports driving through the passage at the crater rim and descent in the direction of the coastal region. Continuing methane rain.

15:17 CHARON encounters a methane stream and named it Styx. It seems to be an outflow from CGH. From the discharge rate, the colleagues estimate a precipitation of 0.05 inch per twenty-four hours, an astonishing amount.

15:40   CHARON reports heavy icing up on the caterpillar. Ethane is frozen on the vehicle after it's crossed the river at a shallow place to avoid the polymer mud. The ford is clearly marked with infrared reflector cells. Takeo tries to remove the ice manually from the track chain. He has only partial success, but the continuation of the journey is possible with reduced speed. They follow the right riverbank. The visibility decreases; the infrared vision devices are activated.

15:47   The Titanauts left behind in the CGH, Shadia and Pablo, arrive at ORPHEUS with the fully loaded sledge. They find the cabin badly cooled down and diagnose a serious defect in the onboard nuclear heat supply. Batteries temporarily supply the electricity. Setting up the experimental methane powerplant becomes a matter of survival. Thanks to the precipitation, liquid methane is abundantly available and is filled into containers.

16:02   CHARON has increasing difficulties with polymers. They consist of acetylene and hydrogen cyanide and are presumably the precipitation of the red-orange aerosols floating in Titan's atmosphere. Methane rain washes the microscopically small particles into the lowlands. There they accumulate into gluey, sometimes yard-thick layers. CHARON reports numerous slimy pools and swamps in different orange and red colors.

17:48   ORPHEUS is now supplied with electricity and warmth by the methane power station. The plant burns methane, gathered in a makeshift rain gutter, and with oxygen out of the storage tank. It still rains. Both astronauts withdraw into the warmed-up cabin. Direct communication with CHARON. Incoming pictures can now be sent live on to HTM.

18:55   CHARON reaches the Kraken Mare at Sinus Numinosus. A purple iridescent sea of liquid hydrocarbons! Mainly methane.[1] The rain stopped for a short time. Clouds repeatedly break up, strong wind gusts. Visibility at least three miles over the methane bogs, greenish electrical discharges over the open ocean. High waves and heavy surf. Crystal clear icebergs drift farther out. They are probably composed of ethane ice. Snow-white bizarre water-ice rocks at the shore. Liquid samples show a high ratio of ethane and numerous organic compounds of diacetylene to acetonitrile. A fully automated permanent observation station is installed, cold resistant for at least two years. Many samples of sand, ice, and sea liquid are gathered and stowed away. The team eats and recuperates.

19:13   Dark clouds over ORPHEUS give reason to expect even heavier precipitation. The wind speed increases. Dull noise can be heard, similar to a large waterfall or distant thunder. Both astronauts measure the most important weather parameters and lay down in the cabin to sleep.

19:51   Persistent soiling of the caterpillar with mud clots delay CHARON's departure. Strong wind develops, drifting clouds, but still dry.

20:03   CHARON starts its way back along Styx, the outflow of CGH. In addition to the samples, heavy rocks possibly of volcanic origin are also loaded. The weather has grown worse, strong precipitation sets in. The polymer mud swollen by the rain further obstructs the heavily loaded caterpillar.

One Hour Later in the Chat Room of the Journalists

"The events on Titan were followed closely on board the Hermes Trismegistus," Hoihong reported. "Sheldon noted with dismay the failure of the nuclear energy supply on ORPHEUS, since a repair under the extreme conditions of Titan was out of the question. In this ice desert of -288°F, warmth was the most elementary property depending totally on an efficient energy source."

"At least the makeshift methane station allowed the planned winding-up of the last mission activities," Astraia said.

"Yes," Randall said, with as much reassurance as he could muster. "The Promethean fire seemed to be saved."

"The expedition to the methane ocean was without doubt the most daring activity on the giant moon," Hoihong reflected. "The nervous expectations of the astronauts were mixed with a still nameless fear."

"The continuous analysis of the meteorological data transmitted by the infrared telescope gave rise to true worries for the early afternoon," Hoihong continued. "The previously stable high pressure in the region of the Crater of Good Hope seemed already to drop rapidly from 12:00 UT and give way to an upcoming storm front from the north that constantly increased and caused extremely heavy precipitation of methane and aerosol."

"In the evening," Randall said, picking up the narrative, "the last round of dispute took place between the Hermenauts during the usual algae menu enriched with mushrooms. It was no coincidence that the talks were about the 'last things,' the distant future and end-time. The mood fluctuated between tense expectation and unacknowledged anxiety. The present worries intensified Sheldon's depression that noticeably developed in the last weeks. His dark dream did the rest."

"Perhaps that was one of the reasons that Thomas formulated thoughts during dinner he would have branded as utopian aberrations in a circle of colleagues on Earth," Astraia speculated. "We don't know whether he gave in to his fantasy or only wanted to express himself in the technologically oriented conceptual world of his partner, Sheldon, in order to pull him out from the dangerous allurements of Saturn, the star of melancholy."

Apocalypse between Physics and Theology

THOMAS: Even with all my enthusiasm for extraterrestrial outdoor adventures, I wouldn't like to swap with our comrades on Titan. The cold makes everything more difficult. No human being ever stayed at such a cold place before.

SHELDON: We've always imagined hell as hot, but the record for the lowest temperature of -301°F just transmitted from Sinus Numinosus is every bit as bad.

THOMAS: Worse! In the Middle Ages, people liked to imagine hot and cold places for penalty. In the deepest region of hell, eternal ice prevails. Lucifer is said to be frozen stiff there! On Titan is truly "weeping and gnashing of teeth."[1]

SHELDON: I seriously question whether this Saturn moon is a suitable location for human colonization. Even the bottom of the sea on Earth is easier to populate.

THOMAS: I learned from you astrophysicists that the sun will become brighter in the distant future and heat up the solar system. Titan could still become a possible living space some day in the distant future.

SHELDON: Yes, the sun is getting gradually hotter in the interior.[2] Helium cinder results from the fusion of hydrogen that provides the sun its energy. As a consequence of this change of composition, the pressure in the sun's interior rises and speeds up the fusion process. Since its formation, the sun has gained one-third in luminosity.

THOMAS: What does this mean for the future of our beloved Earth?

SHELDON: Don't worry. It will take still 5.5 billion years until the sun radiates twice as much heat as it does today. As a consequence, Earth will then become hotter. At the latitude of Anchorage, Alaska, temperatures will be like at the equator today in less than a billion years.

THOMAS: And it will be much sooner if humankind continues to add carbon dioxide to Earth's atmosphere by fossil fuels. Glaciers melt, coasts are flooded and submerged, species are destroyed . . . to the point where parts of the planet become, in fact, uninhabitable.

SHELDON: Yes. The larger greenhouse effect already enhances the temperature and will increase until the carbon dioxide level becomes stable. That will occur eventually within a hundred years either by human reason or a worldwide dearth. The warming up by the aging sun over a billion years, however, is inevitable.

THOMAS: Palm trees at the Alaska North Slope! Luckily, we're still far from such scenarios. In any case, sooner or later the environment on Earth will change. But we do have to think about stretching the biosphere out farther than Earth. Life will populate a climatically more favorable region of the solar system in as great a variety as possible. In a distant future, a new ark can fly to Titan to build a community.

SHELDON: Your mythological inclination keeps seducing you to utopian flights of fancy. However, in the battle against the sun's superiority, you've only won an intermediate victory. In less than seven billion years, it will be also too hot on Titan. At that time, the sun will have inflated out to the diameter of today's Earth orbit and shine two thousand times brighter than today. Earth will be slowed in its orbit by the solar gas and sink into the sun in 7.5 billion years.

THOMAS: Then the conquest and settlement of humanity will continue on out to Neptune and Pluto.

SHELDON: A hopeless flight! Soon after that, our star will have no nuclear fuel available. Severe cold follows after the heat. Only ten million years later, the sun will shrink to a sphere the size of Earth. Our sun suffers the fate of all low-mass stars and will become a white dwarf. As a result of the contraction, it will become hot as never before, but because of its small size, it will have only one-thousandth of today's luminosity, and it will slowly cool down. The temperature at Earth's orbit will sink below -364°F. If it survives for some reason, our home planet will be a dead globe without air and water. The atmosphere will have escaped into space together with all the water because of the excess pressure of the evaporating oceans. Finally, the whole solar system will become deeply frozen. I'm afraid there's no long-term chance of survival for Earth's ecosystem.

THOMAS: We're talking billions of years. By then, life will have spread out long ago to other stellar systems.

SHELDON: I envy your optimism in view of our difficulties on Titan. But, alas, it's truly hopeless! In ten thousand billion years, all the hydrogen in the Milky Way galaxy necessary for forming new stars will be used up. Another ten times more years later, the last star will also become a white dwarf. There's no escape because the mass of our galaxy is limited, and therefore the energy and the possibilities for star formation are finite. Ultimately, any prognosis within the framework of physics must be pessimistic. Life as we know it won't last forever.

THOMAS: I reject your pessimism. Think of what can happen in billions of years! The development of the unicellular eukaryotes to multicellular animals took more than three billion years, but it needed only half a billion years from there to *Homo sapiens*. When evolution advances along so rapidly, many things are possible. Life could even break away from carbon. I remember a colleague at the university who enthused about the new possibility for the development of life on the basis of silicon.

SHELDON: Until now, the evolution of life has taken place in the world of organic chemistry and can't jump over to silicon. It crosses my mind, however, that the designers of modern computers learned from biological evolution and used the same principles. In the late twentieth century, computers were designed and built like machines, but now, in the middle of the twenty-first century, engineers have begun to develop them according to mutation and selection. A huge number of devices are produced with slightly different properties. The best is selected, the others recycled. This second evolutionary line is based, however, on other substrata than biology. The material bases are totally different and will mix as little as oil and water.

THOMAS: Some futurologists expect a total transfer of chemical information of the human brain to digital data storage media and computer systems. The other body functions, though, must be simulated, such as releasing messenger substances and hormones. The human consciousness could then be embodied on the most advantageous matter in order to continue to exist eternally in low temperatures and with a minor need of energy.[3] It would also be conceivable that different intelligences would interlink and integrate into a larger consciousness.

SHELDON: Copy that. You're thinking of some kind of supermachine intelligence.

THOMAS: I am not talking about present artificial intelligence, which is a fast but poor emulation of natural intelligence. I imagine a development of intelligence comparable to the evolutionary step from single-celled to multicellular organisms, when individual cells differentiated to operate as a whole and completely new functions emerged.

SHELDON: You're very good in cybernetic speculations, Thomas. It's possible that the development of information technology will some day overtake biological evolution, since computer evolution won't be driven by chance but by humans.

THOMAS: Of course, I mistrust these utopias, which consider everything to be technically feasible. In any case, a life of millions of years on a silicon basis with metabolism reduced to practically nothing is not one of my highest expectations for happiness. Let's take this a little further. What else do we have to expect in the distant future of the universe?

SHELDON: Provided that we'll still be on Earth, we may protect ourselves against dangerous asteroids and comets. But sometime in the next 10^{15} to 10^{17} years, a star will come so close that the planets will be torn out of their orbit around the sun and get lost in outer space. We humans will always be powerless against stars.

THOMAS: Perhaps then we will have long left our solar system.

SHELDON: Encounters between stars radiate gravitational waves and give off energy. Therefore, our galaxy continually shrinks. The stars come ever closer to the center where a gigantic black hole eagerly awaits them. In 10^{19} years, this abyss will swallow most of the stars.

THOMAS: But some of these black holes may, perhaps, become refreshing oases of life in the midst of the dying stars, since near black holes, free energy will still be found in this distant time.

SHELDON: One could live more or less comfortably in the center of galactic black holes, but their energy would also be exhausted after 10^{100} years. Then it would explode into a fireball of energetic particles and gamma quanta. Perhaps the atomic nuclei decay even earlier. Protons probably don't exist forever, but they have a half-life of some 10^{31} years. Their disintegration will drag all known forms of matter with them into extinction. In this most distant future, the universe will consist of only a very thin gas of electrons, positrons, photons, and neutrinos.

THOMAS: Will this be the open end of the gloomy cosmic drama?

SHELDON: It depends on whether enough matter and energy are available in the universe to turn the expansion around. But it is the strong consensus among cosmologists today, since the remote galaxies are moving away from us—the farther, the faster. There are no signs of the least deceleration. On the contrary, the expansion seems even to accelerate. Yet the heated discussions have never died out. Maybe you theologians know more and don't tell us.[4]

THOMAS: We've always been regarded as specialists for eschatology, the "teaching of the last things." In fact, many stages of your scenario remind me of old apocalyptic images. Sun and moon will darken, stars will fall from the sky, and heavenly powers will be shaken.[5] The Book of Revelation by John works with this material, declaring that the whole creation will be afflicted by a cosmic catastrophe.

SHELDON: I don't suppose that you theologians want to co-opt scientific predictions.

THOMAS: Heaven forbid! The timescales alone are different beyond imagination. You reckon with inconceivable periods of time. Compared to them, even the total previous life span of the universe is ridiculously small.

SHELDON: Indeed! Physical eschatology can leave us cold. Yet, a nightmare troubled me last night. I saw the sun in the far distance change into a red giant in time-lapse. The night creates a thousand monsters!

THOMAS: Yes, I know. Have you forgotten that you told me about it this morning? Fears of this kind troubled people in the past far more. They expected enormous catastrophes in the immediate future, perhaps even tomorrow or the day after. This was the case when Jesus lived. Expectations about an end-time haunted people in many epochs.

SHELDON: My prognoses into the distant future aren't born out of fears but are based on rational scientific extrapolations of physics.

THOMAS: I share your aversion to sects. I esteem the old Jewish apocalypticism at the time of Jesus much higher than today's groups gripped by the doomsday fever. Many apocalyptic prophets at that time were learned people, even wise men, who tried to associate the old Israelite traditions with the priestly and natural sciences of their day. They influenced Jesus and Paul. The apocalypticists fostered a "knowledge of the Most High"[6] with enormous theological effort. They continued in their way the integration of theology and cosmology that, centuries before, the priestly writers had provided in the first biblical creation narrative.

SHELDON: Didn't fear play a totally dominant role with the apocalypticists? The imagery in John's Revelation is awfully frightening.

THOMAS: Indeed, fears have a part in the forming of apocalyptic texts, particularly since they were written in situations of extreme distress caused by the world powers at that time. As astonishing as it may sound, the apocalyptic prophets aimed at freeing their followers from fear and despair. They wanted these apocalyptic texts to have a cathartic effect; they allow orientation in an otherwise incomprehensible and hostile world.

SHELDON: How should that happen? How can horror images inspire confidence and encourage action?

"The conversation was interrupted by a news update from Titan," Hoihong reported. "The outlandish pictures from the Sinus Numinosus, which were transmitted via CHARON, made the hearts beat faster also on the mother ship and evidently banished for some time the anxious thoughts and oppressing worries."

"But wait," Astraia exclaimed. "Information from the infrared telescope on Titan's troposphere made it quite clear that the Titanauts trio would have to face a more difficult return trip to the warm ORPHEUS in the Crater of Good Hope."

"We'd better keep listening," Randall said.

SHELDON: Look at this tremendous view over the billowy methane sea. The weather has only apparently calmed down; the forecast is still bad. With that, we're back to our topic: images of the end-time provoke dread and horror.

THOMAS: The description of forthcoming catastrophes is not the main goal of apocalyptic proclamation. The horrors of the end-time are only the deep shadows that the coming light casts: the completion of God's work of creation. Jesus relied on God revealing himself soon in all his fullness. The Revelation written by John places this great future salvation in the center: a change of the whole cosmos into light and spirit, and the unification of all living creatures with the divine ultimate ground of being.

Fantastic images of a new heaven and a new Earth, of the new Jerusalem, where God will dwell in the midst of his creatures and wipe away all tears, because "death will be no more."[7]

SHELDON: Science has nothing to report about such things. Our distant prognoses are cold and pessimistic. In the end, we can only forecast with certainty the decay. It's a mystery to me where you get your confidence. Your prophets awaited a cosmic catastrophe in the near future of their time. We are two living examples that they have been wrong. Jesus was also mistaken. Why do you hang on to such a failed hypothesis?

THOMAS: I can't reproduce the cosmological ideas of antiquity. When I talk about a completion that encompasses the whole creation, I don't mean a scientific hypothesis but rather an elementary hope, and not only for myself. It views the world as a whole.

SHELDON: Modern cosmology is about the known past and present, but not about the unknown future.

THOMAS: That's right. Here, anticipation comes into play. For Jesus and the early Christians, the future completion already radiated into the present. They already felt what will come in abundance over the whole creation. Perhaps you remember our discussion on Easter Sunday.

SHELDON: The step from such conditions to predictions about the future of the universe, though, isn't plausible.

THOMAS: Statements of hope are no prognoses. Whoever gets to feel a touch of divine presence here and now, trusts that this creative power can also change the whole cosmos. The source of the new won't cease to flow.

SHELDON: Physical forecasts promise nothing good. The decay of matter and its structures can't be stopped.

THOMAS: Jesus made use of vivid pictures and parables without offering any exact predictions. There is a remarkable saying: "The kingdom of God is not coming with things that can be observed [by means of astronomy];

nor will they say, 'Look, here it is!' or 'There it is!' For, in fact, the kingdom of God is among you!"[8] Jesus encourages us to be surprised by the coming God. He calls for *alertness*, for attentiveness for that which will come and is already present. Beware, keep alert; for you do not know when the time will come.[9]

SHELDON: To be awake has its time and sleep has its time. We'd better allow ourselves some hours of rest. We don't know what news will reach us from Titan during the next hours. I don't like the storm over the Crater of Good Hope. We should be ready tomorrow morning to assist our friends in an early start and guide them out of this witches' kitchen.

"A little later," Randall told his colleagues in the chat room, "the following alarming news reached the mother ship. Here's what the logbook says."

21:14 Accident! CHARON slides at a traverse on a slippery slope of an ice hill near the coast and gets stuck in a polymer swamp. The astronauts jump off in time. Takeo injures his arm. His heated insulation clothing is ripped and frozen ethane snow sticks on the skin. The rover sinks deeper and deeper. Sergei and Nicole are able to save a large oxygen container and some water supply.

21:31 CHARON is lost! The vehicle has totally disappeared in the mud and cannot be saved despite heroic attempts. Takeo's arm is cleaned and bandaged with remote advice from Paolo, the physician. At the beginning, Takeo felt little of his injury, but now the wound begins to hurt.

21:51 CHARON's crew prepares for the march back on foot. The distance to ORPHEUS in the crater is over nine miles. Fortunately, Takeo was able to save the mobile communication unit, plus the parabolic aerial from the CHARON before it sank into the swamp. The connection with ORPHEUS and *Hermes* remains, but it is greatly reduced to save energy.

"The accident was noted on the spaceships with great consternation," Randall explained.

"But from an engineering point of view," Hoihong replied, "there was no reason for overly dramatizing the situation, since there was still a good thirty-two hours to go until dusk. The CHARON team had with them enough energy, oxygen, food, and water for over twenty-four hours in sufficient, highly concentrated form. The hostile landscape, the adverse weather, and the bulky suits made movement difficult, but the low gravity allowed wide, floating jumps over obstacles."

"The Titanauts were already well trained in stabilizing their balance in difficult terrain," Astraia agreed. "The infrared reflectors the CHARON had laid on its outbound journey marked the trace like a red thread so that the returning astronauts could orient themselves in the dark world of fog patches."

23:34   CHARON team is assisted by tailwind but has trouble with visibility.

02:33   CHARON group reports strong methane rain. They are tired and advancing only very slowly.

02:56   The ORPHEUS crew hears clear sounds of thunder and observes occasional deep-red fluorescence over one-third of the sky. Gigantic flashes of violet lightning come down at the western horizon. A thunderstorm seems to rage over the CGH.

06:45   The methane river Styx has greatly swollen. The CHARON colleagues see no possibility of wading across at the marked ford. They wait at the bank and lay down exhausted to recuperate. The distance to the crater passage is now only a little less than two miles, but uphill.

09:09   The ORPHEUS crew considers driving to the ford with a transport sledge and pioneer bridge to make a crossing possible.

09:20   ORPHEUS is shaken by an explosion. Perhaps an atmospheric discharge ignited the oxygen-methane mix. The new power station breaks down. ORPHEUS gets its energy again from the batteries. The emergency power unit consisting of fuel elements is too cold and cannot be started up. The explosion seems to have done still more damage. The ORPHEUS crew tries desperately to fix the damage. The communication with HTM is severely hampered.

"The batteries consume electricity for their own heating and last only for thirteen hours," Hoihong explained. "During this time, ORPHEUS should takeoff. Without electricity, the computers cannot operate and the booster rockets of the ascending module cannot be ignited. If they won't start, the crew could survive for another thirty hours. There remain about twenty hours until dusk and another eight hours to total darkness."

11:53   The ocean expedition reports in again. The river's methane level remains high. The three Titanauts decide to take a different route for their return. In order to conquer the jagged, steep ice rocks of the mountains at the rim of the crater, they look for a pathway in the area of the southern volcano. The oxygen will last for at least ten hours.

15:43   Decreasing cabin temperatures are reported from ORPHEUS. The batteries' voltage drops since they are inadequately heated. The methane power station cannot be repaired. The emergency power unit remains dead.

16:02   The colleagues from CHARON make good progress, but they calculate that it will still take two hours to the volcano. White frost of ethane ice covers the aerial. That is probably the reason the connection becomes weaker. They don't know how long the batteries of their transmitter will last.

16:57   The signals from ORPHEUS also become weaker. The energy supply is deficient. The Titanauts reduce the cabin heating and switch off unnecessary computers.

17:34   All attempts for repairing the methane power station and the emergency power unit are futile. The ORPHEUS people are worn out and retreat into the cabin. They keep their isolation suits on to use the energy reserves optimally.

18:17   The ocean expedition reaches the volcano.

18:24   ORPHEUS is barely audible. The colleague at the radio speaks very slowly. It must be very cold in the cabin.

18:29   The radio connection between ORPHEUS and HTM is lost.

After some moments of silence in the chat room, Randall Bradford reported about his inquiries in the early Australian morning:

> The increasingly meager information that reached *Hermes Trismegistus* from Titan in the course of May 20 understandably got under the astronauts' skin. Remember that the disaster on Titan already was looming with the failure of the primary nuclear energy supply the previous day. As lively the interest Sheldon and Thomas had shown in the incidents down below, they couldn't see any way to help from their position. The horrible thought to have to leave their colleagues behind on this cold moon remained for the moment unspoken, but presumably deeply stirred up their emotions. Nagging feelings of dread and abandonment seemed to alternate with phases of new hope and a desperate search for a solution.
>
>   The enforced inactivity must have intensified Sheldon's depression to the point where he couldn't chase it away anymore by routine work. Nevertheless, it's certain that in the morning he discussed with Thomas the possibility of flying a rescue operation with the second smaller landing craft still docked at the mother ship. Although this craft was primarily slated for a manned

mission to the moon Iapetus, it could also be used for landing on a celestial body with atmosphere, if necessary, thanks to IASA's famous FAME (Flexible and Modular Engineering) technology. However, the cosmonauts would have to take no small risk with a landing maneuver on the thickly clouded Titan.

At least the ORPHEUS team had already mapped the target zone and not only calculated the optimal approach path but also tested it successfully. In view of the towering difficulties of the Titanauts, the challenging rescue project took on a more and more concrete form in the course of the afternoon. It's also likely that the prospect of a specific action helped Sheldon banish the dark clouds of his dejection.

We know a little more about how Thomas thought. Like Sheldon, he had an extremely troubled night. His diary drips with old mythology, but during the night, Thomas also rehearsed in his mind the technical possibility of a rescue trip. Thomas identified himself with the Titanauts, above all with the team of three on their erratic track through the shadow world of the giant moon. This makes it understandable that he searched for a way to save them. We found his computer logs estimating the amount of fuel for an ascent from Titan's surface. Science and religion were no longer clearly separated but complemented each other for valuation and decisions. This is his last entry in his diary:

Descending into Hell
From Thomas's Diary (May 19–20)

My exultant joy at the sight of the "holy sea," the crimson Sinus Numinosus of Titan, is suddenly gone. The lofty display is replaced with blunt fear. The anxiety regarding our friends in the world of shadows holds me with an iron grip. We receive some information from the CHARON team only every few hours. My mind searches for rest, but the sleep will not stay. In the midst of the approaching waves of slumber, my restless mind gropes for an uncertain future.

In the twilight zone in the deepest circle of the underworld of my consciousness, a dreadful image of Cronus emerges. The mighty god has four eyes—two in the front and two in the back; two of them focus sharply while the other two are closed. On his shoulders he has four wings; two of them flap and two hang down. An old oracle voice resounds like thunder: "Cronus was watchful even when in repose, and was in repose even when awake; he flew while at rest, and was at rest while flying."[10]

Startled, I chase away the nightmares and float to the cabin window. Saturn shows me its dark countenance averted from the sun. The gigantic round body covered in dark gray looks unfriendly and unapproachable in front of the deep black universe. Recollections of ancient myths haunt me. Outside hangs the age-old god in space, incarcerated in the dark Tartarus, bound by his son Zeus with untearable chains. Hesiod's verses paint the somberness of that location with frightful colors:[11]

> *That is where the Titan gods are hidden*
> *under murky gloom by the plans of the cloud-gatherer Zeus,*
> *in a dark place, at the farthest part of huge earth.*
> *They cannot get out . . . That is where the sources and limits of the dark*
> *earth are, out of murky Tartarus, of the barren sea,*
> *and of the starry sky, of everything, one after another, distressful,*
> *dank, things which even the gods hate: a great chasm . . .*
> *it is terrible for the immortal gods as well, this monstrosity;*
> *and the terrible houses of dark Night stand here,*
> *shrouded in black clouds.*

And still, there were heroes who left the sweet light of the day behind them and climbed down into the gaping underworld: Orpheus and Heracles, Odysseus and Aeneas. Barely a one reached the abyss of the dark Tartarus—except Dante and his guide Virgil.

Nevertheless, the ancient Jewish visionary Enoch is said to have been sent from heaven into the underground gorges.[12] *There, namely, was the prison of the lofty angels, who once before the Flood, driven by fatal desire, seduced the beautiful daughters of men and brought magical arts to Earth.*

Enoch was carried away. He not only explored Paradise and all four directions of Earth, but also this location of the fallen. He reached the "deep pit with the pillars of heavenly fire," and on the other side he "saw a place without the heavenly firmament above it or earthly foundation under it . . . a desolate and terrible place." And he "saw there the seven stars which were like great burning mountains"—the fallen star angels, who at the dawn of creation rebelled against God. The accompanying angel Uriel declared to the trembling Enoch that these angels remain bound here for all eternity.

A shiver runs through me in view of the abyss, opening into a bottomless chasm. An icy breeze of forsakenness wafts toward me and freezes my heart. This is the place in hell that's most distant from God in earlier time. Burning pain and raging rebellion have faded away in this innermost sphere of the underworld. Left is only irrevocable loneliness and despair, silent separation from all life, a time frozen to a terrible eternity . . .

A Blinding Beacon in Darkest Night

But the living God came down even into this abyss. An image of hope enters my fainting heart like a hot spring: the Son of God in the depths of hell.

Since early times, Christians have talked about how Jesus went down into the underworld between Good Friday and Easter morning and broke open the iron gates of the realm of the dead to bring rescue and salvation to the dead, to the first parents, Adam and Eve, along with all of their pious descendants. The risen Christ looks triumphantly back on his voyage to hell and meeting with death:

> *The underworld saw me and was shattered,*
> *and Death ejected me and many with me.*

I have been vinegar and bitterness to it,
and I went down with it as far as its depth.
Then the feet and the head it released,
because it was not able to endure my face.[13]

"At this point, Thomas assembled fragments from New Testament apocrypha," Astraia injected. "I know them from my studies in the British Museum."

Love led Christ down as far as to the dead, who had given up all hope, separated from the stream of life.

And I made a congregation of living among his dead;
and I spoke with them by living lips ...
I placed my name upon their head,
because they are free and they are mine.

Figure 11. *Descent of Christ to Limbo.* (Andrea di Bonaiuto, Spanish Chapel, 1365 to 1367, Florence, Italy /Peter Barritt/Alamy Stock Photo)

Christ saves the dead and leads them up to the light of heaven. The gates of Hades are unlocked forever.

The metaphor of Christ's trip to hell widens in my mind in dizzying dimensions. The Son of God, nailed on the cross, gave himself to the devouring nothingness of death and penetrates all existence with his creative love. His living light catches up with the expanding space and the passing time. There is no depth, no matter how far from God, which he does not fill with his saving presence, no erring being he will not bring home at last—until hell is completely empty and a memorial of past suffering, sealed with the sign of the cross. Thus the Deity finally reconciles the whole universe, changed it, and takes it into his holy light . . .

Mysteries of the Underworld

My mind returns back from the overwhelming light of the resurrection world into my small cabin. The thought about our colleagues returns again like a painful incision. For more than three years, we shared with them our accommodating Hermes *in the cosmic loneliness, and finally reached our destination in the mysterious world of the fallen old gods. But our friends dared to go further. In the depths of Titan's shadow world, they came across a source of life and looked at the holy sea. The promise of the guide to Hades, Orpheus, rings in my ears:*[14]

> *They will give you water to drink from the lake of Remembrance,*
> *So that, once you have drunk, you too will go along the sacred way*
> *by which the insiders and adepts advance, glorious.*

The happy inebriation does not last. Our friends did not find beatific consecrations down there and no holy road that could bring them up again into the sphere of light. Now they roam around in the dark depths, left alone and threatened by the icy cold, the storm, and will-o'-the-wisp of methane swamps. Powerless, we have to wait for them up here and protect the weak flickering candlelight of hope from Titan's unleashed forces.

The thought of the Son of God in the realm of the dead infuses me with new courage. My soul, almost a shadow being of Hades, fills with invigorating power, as if it would be infused with honey and blood. There is a saving way down into the dark halls of Hades, past the threatening monsters, to assist our friends.

A small shining crescent shape shows over the rim of Saturn's dark round body—the sign of the rising distant sun. Determined, I get up. The courage of a bold warrior enters me. Images of the secret mysteries of Mithras appear before my mind's eye. It is the consecration path of the warriors for the good.

I stride through their seven consecration levels, corresponding to the planet's spheres—I become raven and groom, soldier and lion, Persian and sun-runner. Unflinchingly I reach for the insignia of the seventh level, the sickle, the rudder, and the staff. Under the sign of Saturn, the highest sphere, I change into the fiery pater Mithras. Courageously I turn to the ancient gates, the gaping underworld, to bring help to our friends.

The Outcome of the Saturn Mission, or a Space Odyssey 2051

"Good morning, my friends," Randall started the discussion from Australia. "While you were still sleeping, I wrote a draft about the last hours of the mission to Saturn for Harold and the Board of the International Aeronautical and Space Administration. Here it is:"

The dramatic deterioration of the events understandably left no time for the two astronauts on board the *Hermes Trismegistus* to make extensive entries into their logbook or even their diaries. Nevertheless, the last fifteen hours can be reconstructed in detail. Since the media previously reported in depth about the final phase of the Saturn mission, we limit ourselves here to the most important incidents and try to examine the few occurrences where the astronauts seem to interpret the circumstances that have become difficult. We take the liberty of posing the question whether, consciously or unconsciously to a large degree, they tried to carry out insights of their previous discussion topics.

Based on the catastrophic signals from the ORPHEUS crew, the astronaut team made the momentous decision for a rescue operation on Saturday, May 20, at 18:48 UT. The control station on Earth was only informed very late, apparently to avoid lengthy discussions. All investigative commissions of inquiry set up later by IASA have explicitly sanctioned the undertaking that may appear panicky at first glance. But a fast intervention seemed to be the only way to save the crew of the ORPHEUS.

The speculation made by the press remains unchallenged that the unauthorized decision by the two astronauts of *Hermes Trismegistus* was opportune as far as IASA was concerned, so they would not be made responsible for the outcome.

Subsequently the second, smaller landing craft, the Emergency Lander and Prospector for Ice-Moon Studies (ELPís, significantly the Greek word for "hope"), was made ready for operation. Since both astronauts still considered the possibility to repair the energy supply for ORPHEUS, they loaded extensive repair and replacement material into the landing craft's cargo space. At the same time, the necessary trajectory computation for the mission had to be made and copied into the ELPís's central computer. Additional oxygen and food supplies were also to be thought of, as well as fuel tanks and medical material.

Almost no personal references from Sheldon Cutter are left behind about the last hours on board the *Hermes Trismegistus*. The main responsibility of the complex operational planning was solely on his shoulders. It seems clear now that the initial frantic activities dispelled his prior depression and that he found the way back to his robust mental constitution.

As an experienced astronaut and former rescue pilot, he was trained to handle difficult situations, and the available means of the most modern technology seem to infuse him with new confidence. Above all, the relationship with the ORPHEUS crew that had grown over the many years of teamwork gave him full access to his resources. The voice recorder taped only the suggestive saying he pressed out on several occasions, which seemed to accompany him during the preparations:

"Do what you believe in, and believe in what you do."

It was different for Thomas Haubensak. Within a few hours, he had to become acquainted with the programs for orbit calculations of spaceships. He too found no time for writing, but he had many monologues talking to himself in an undertone. Unfortunately they were recorded only in fragmentary condition by the data storage. His inclination for exaltation seems to have been rather increased by the escalation of the events, although an emotional episode, diagnosed by some experts as psychotic, was in no way supported by the available facts.

Understandably, Thomas developed further his thoughts noted down in his last diary entry. They revolved around mythological underworld voyages.

An enigmatic saying often recurred. Experts identified it as a saying from the ancient Greek philosopher Heraclitus:[1]

"The way up and the way down is one and the same!"

The commemoration of Ascension Day, ten days ago, had powerfully stimulated Thomas's already overexcited fantasy. He interpreted the planned landing on Titan as an underworld voyage, an ascension in reverse, so to speak. He imagined again the distinguished set of figures from antiquity known for their daring trips to Hades, especially Orpheus and Odysseus.

Nevertheless, the goal of rescuing the colleagues was strongly imprinted on his mind. In the public debates of a few days ago, it was frequently overlooked how strongly a group of people, living together over several years and almost totally isolated from outside contacts, is forged together. It's remarkable and admirable that this team of seven astronauts, despite occasionally explosive erupting tensions, worked things out with one another many times in a community of fate and partnership. These strong personal bonds became also for Thomas an all-domineering motivation for the rescue mission.

The contrasting expert opinion of the psychiatrist Titus Hornschuh—controversial in other ways as well—proposes that Thomas identified with Jesus Christ, who went down into the underworld on Easter Saturday to save his ancestors Adam and Eve together with all of their offspring.

We consider this model of interpretation incorrect. Thomas had to wrestle with completely different problems during the last known hours. Apart from that, it was very clear to him that his descent into the shadow world of Titan should be directly before Pentecost, not like Jesus in the Easter night.

The pair of astronauts started on May 20 at 22:45 to their own unforeseen Titan mission. As before, the position of the mother ship *Hermes Trismegistus* was monitored and controlled automatically at the Lagrange point between Saturn and Titan. We know that the large spaceship is still at this location today, as if it awaited loyally and unperturbed for the return of the missing Titanauts.

During the flight, Thomas calculated the flight path of ELPÍS, in constant feedback with the powerful central computer on *Hermes*. It also stored the extensive Titan mapping data by the ORPHEUS crew. After a good five-hour flight, the landing craft turned into an orbit around the gigantic moon.

The asymmetry in the allocation of duties changed notably between Sheldon and Thomas during this approach maneuver. Sheldon was totally engaged with the manual control of ELPÍS. The duty of the continuous recalculation of the course and the input of the latest data from the board radar rested with Thomas. Some of his comments clearly indicate that the forced takeover of larger responsibilities due to the emergency intensified the feeling of equal partnership with Sheldon.

The tense situation hardly allowed an exchange of feelings and anxieties of both astronauts. The fervent wish that everything may turn out well united them far more than all earlier extensive discussions about everything and anything. We don't know if each, quietly on his own, somehow hoped for a helping power—Thomas, his saving God; Sheldon, perhaps for a favor of cosmic chance or the ever-new-creating force of evolution. But it's clear that what unified them as never before was the decisive dedication for jointly assisting the endangered friends in the ice desert of Titan.

At 04:47 on May 21, ELPÍS approached Saturn's moon to 165 miles and started with the descent. Even though Thomas was fully occupied with the calculations for an optimal landing trajectory, he let himself be carried away occasionally at the sight of gigantic reddishly shimmering celestial body and burst out in astonishment.

At the altitude of 137 miles, Sheldon began with the entrance maneuver into Titan's atmosphere. Some rather unusual turbulence at the altitude of thirty-one miles, recognizable from high above, forced him to aim for a relatively flat approach path to the Crater of Good Hope, where ORPHEUS had landed. Under these conditions, the forthcoming maneuver would drag on for more than four hours. The trajectory computations required highest precision to avoid the dangers of burning up or bouncing off the moon's dense atmosphere.

During the following hours, relatively little was spoken. The data storage of *Hermes Trismegistus* contains very detailed information about the first phase of the landing that initially went according to plan. Since ELPÍS's deceleration stayed somewhat behind the calculations, Sheldon swung it into a spacious loop.

At 06:02, they flew over the destination at an altitude of 112 miles at Mach 1.5. The region was still under the influence of the extensive storm that had already badly afflicted the ORPHEUS team. Sheldon was determined to take no risk and chose the flattest path to the crater. A further orbit of the moon proved to be necessary. At an altitude of 108 miles, he opened the small pilot parachute, and thirty seconds later the main parachute, to reduce the still high speed. Thomas transmitted the course data to *Hermes* at 07:37 at an altitude of seventy-four miles for the last time.

The last hour and a half present us with many puzzles. Apparently, the complicated evasive maneuvers of ELPÍS led to considerable navigational problems. The enormous electric fields produced by the violent storms may have increasingly disturbed the flow of information from ELPÍS to *Hermes* and back. The exact flight path of the landing craft cannot be reconstructed precisely.

At sixty-three miles altitude, the large parachute was dumped according to plan to make room for the third parachute that stabilized and slowed down the steep gliding flight.

At fifty miles, they penetrated the thick condensate haze layer. Sheldon had to risk entering into the target area stricken by storm turbulences and to accept considerable orbital path irritations, additionally intensified by the parachute. The orientation seems to have become increasingly difficult for the team, although after cutting through the cloud layers, they were not forced to blind flight, as often claimed in the news coverage two weeks ago.

Sydney, June 20, 2051

"This is how far I got," Randall said. "What about the last ten minutes?"

"The outcome is clear," Hoihong insisted. "As an engineer, I propose to understand the words of the astronauts literally as expressions of a reality they observed. Why not accept their words? They must have encountered living beings."

"No," Astraia objected. "The last words seem to me to express more than simple sightings by eyes. What they saw may have been more like visions. These words remind me of the first dialogue that the two astronauts conducted on Easter morning. There, the question was left open whether the Easter witnesses had seen material events, visions, or a spiritual reality perceived by participation. The same possibilities—objective facts, visions, or spiritual reality—also hold in view of the last descriptions of the astronauts."

"There are striking similarities between Kubrick's *2001: A Space Odyssey* and the last communications," Randall pointed out. "The movie is open-ended and leaves an unresolvable mystery, appropriate to the inevitable enigmas of the universe. Here, too, we may have our own interpretation, but the question must remain open."

From 08:12 to 09:04, the *Hermes Trismegistus*'s computer memory contains only fragmentary information from the speech recorder in ELPÍS.

Randall added: "Here is the transcript they left. Editorial notes are set in italics. OMEGA means the target region, the Crater of Good Hope."

08:14	**Thomas:**	Altitude 25 miles. At all costs, keep the present approach angle.
08:15	**Sheldon:**	Massive methane clouds at 19 miles! That is the storm zone. We are still too fast for penetration. Flying another loop.
08:20	**Thomas:**	Humongous, these scarlet and orange fluorescent cloud giants. Truly titanic! Do you see the deep-blue lightning activity in the inside?
08:20	**Sheldon:**	It must be extremely uncomfortable down there. I'll try to come in from the west.
08:26	**Thomas:**	Please don't swerve out farther. The distance to OMEGA is increasing rapidly.

08:27	**Sheldon:**	OK, now we enter flat into the cloud layer. Down into the hell of a mess. Hold on tight!
08:30	**Sheldon:**	(*after unintelligible words, covered up by noise*) Oh man! The parachute increases the rocking.
08:30	**Thomas:**	Just had the first view of the ground. Altitude 11 miles.
08:30	**Sheldon:**	A massive colossus in front of us. Can't dodge it. (*strong noise builds up*)
08:40	**Thomas:**	Navigator's acting up. We've lost our course!
08:40	**Sheldon:**	We'll be below the middle cloud layer at any moment.
08:41	**Thomas:**	Heavy methane rain. I don't know where we are. (*from now on, a constant background noise interferes with the data transfer*)
08:42	**Thomas:**	(*after unintelligible words*) Altitude 5 miles. Navigator operates again, but ridiculous numbers.
08:43	**Sheldon:**	Again, a turbulent but thin layer of clouds. Annoying, this ice that built up on the left porthole.
08:43	**Thomas:**	I'll try to increase the window heating. There, again a view down below. I see a jagged mountain range.
08:44	**Sheldon:**	Strange light appears to the right. Those must be ice crystals in the air.
08:45	**Thomas:**	Monstrous; in front of us a gigantic rainbow. Wow! Brilliant. Fluorescing all around. Behind us, the clouds are breaking up. I can hardly believe it. The sun shines in directly from the back.

08:47	**Sheldon:**	(*after violent noise*) No, it can't be. I see Saturn on our back monitor. Can you see it too?
08:47	**Thomas:**	Watch out. There in front is a large, dark cloud bank. Can you get around it? (*unintelligible words covered by noise*)
08:48	**Sheldon:**	(*after heavy noise dies down*) I can't go lower, as long as we're not sure we're at OMEGA. Gliding again extremely flat.
08:50	**Thomas:**	Again the deep-red mountain. What's behind it?
08:50	**Sheldon:**	The crater! It's the crater! OMEGA!
08:51	**Thomas:**	Hopefully the right one! Go lower.
08:52	**Thomas:**	Altitude 9,200 feet. The navigator's acting up again!
08:53	**Sheldon:**	Do you see the rugged mountains in front of us? Dark red with enormous snow-covered peaks. Like a gigantic saw. On the left, a crater.
08:53	**Thomas:**	Patches of methane fog. Terrible ghostly lights. Where are we?
08:57	**Sheldon:**	We're now flying over the crater rim. Unbelievable ice structures. Sharp like knives. Looks like a titanic palace made from ice.
08:57	**Thomas:**	(*after some unintelligible words*) There in front, again deep red, at the crater wall! A volcano, steam is coming up. Go closer! Perhaps we're right after all. (*howling of a storm begins, shredding the sentences into fragments*) Strange glaring light! From where . . .

08:58	**Sheldon:**	The volcano. Hot springs steaming . . . A towering geyser under us . . . There is something! There are structures at its foot . . . immense. Incredible!
08:59	**Thomas:**	Colossal structures . . . Look organic. Heavens, like titanic mushroom formations! That can't be!
08:59	**Sheldon:**	That is life! It pulsates . . . light effects! It lives!
09:00	**Thomas:**	Do you feel it too? . . . It lives! It thinks! I can feel its dreams!

Elpís's chronometer fails to work. The following three minutes are calculated back from the received data on *Hermes*. The howling of the storm suddenly stops; it becomes quiet, only a quiet, soft whistle. The last word fragments are totally clear but disconnected.

09:01	**Thomas:**	Contact! Something is making contact with us . . .
09:01	**Sheldon:**	We're being received. Empathy, warmth . . . Something good, endlessly good . . .
09:02	**Thomas:**	I sense its dreams . . . It dreams about the universe . . . The universe . . . it is all one! . . . The entire universe—it lives . . . it breathes. It is alive!
09:03	**Sheldon:**	Our friends from ORPHEUS and CHARON . . . ! I can feel them . . . They are here, secure . . . Us too . . . Together with them—SAVED! SAVED! SAVED!
09:04		(*Connection terminated*)

Abbreviations and Acronyms

ARGOS   **A**rrayed **R**eflector **G**adget for **O**ptical **S**tudies, a one-hundred-foot telescope on *Hermes Trismegistus*. Argos is also the Greek name of the mythological Argus, an "all-seeing" monster with one hundred eyes. He has been killed by Hermes.

CHARON   **C**arrier for **H**umans and **R**etrieval of **O**bjects of **N**ewness, a tracked vehicle (caterpillar) on the surface of Saturn's moon Titan. Charon is also the mythological ferryman who carried the shades of the dead in his boat across the river to their final abode in the underworld.

CGH   Crater of Good Hope, the landing location of ORPHEUS on Titan.

EDEN   **E**xperimental **D**ivision for **E**cology and **N**utrition, the bio-zone of the spaceship. In the Bible, paradise is called the "Garden of Eden."

ELPÍS   **E**mergency **L**ander and **P**rospector for **I**ce Moon **S**tudies, a manned landing craft, originally planned for Saturn's moon Iapetus. *Elpis* is also the Greek word for "hope."

HTM   Human Transfer Module; alias *Hermes Trismegistus*, the mother ship of the *Hermes* mission. In antiquity Hermes, the Greek god of communication (called *Mercurius* by the Romans) was combined with Thoth, the Egyptian god of wisdom. The surname Trismegistus means the "threefold largest," namely the largest philosopher (scientist), priest, and king.

IASA   International Aeronautical and Space Administration. Fictitious space agency.

ODYSSEUS **O**rbiting **D**evice for **Y**earlong **S**aturnian **S**ystem **E**xploration and **U**nrestrained **S**tudies, an unmanned exploration probe to Saturn's smaller moons. Odysseus is also the hero of Homer's epic poem *The Odyssey*. He strayed the Mediterranean for ten years to get home after the Trojan War.

ORPHEUS **O**rbiting **R**adar **P**latform and **H**abitat for **E**xcursions to **U**nknown **S**atellites, the landing craft of the crew to Titan. In Greek mythology, Orpheus is the supreme singer and musician who descended to the underworld to bring back his deceased wife.

PAN **P**enetrating **A**utomatic **N**avigator, an unmanned ring probe. In Greek mythology, Pan is a rural god of shepherds and flocks, part man and part goat, whose father was Hermes.

SETI Search for Extraterrestrial Intelligence, a recognized field of astronomy.

SMART **S**elf-**M**aneuvering **a**nd **R**obotic **T**ransporter, simple unmanned supply ferries.

UT Universal Time or Greenwich Mean Time.

Notes

Scripture quotations are from the New Revised Standard Version Bible, copyright © 1989 the Division of Christian Education of the National Council of the Churches of Christ in the United States of America. Used by permission. All rights reserved.

Crisis on the Elpis

1. Rosaly M. C. Lopes, Stephen D. Wall, Charles Elachi, et al. *Titan as Revealed by the Cassini Radar*, Space Science Review 215 (2019):33. Jonathan I. Lunine, Ralph D. Lorenz, and William K. Hartmann, "Some Speculations on Titan's Past, Present and Future," *Planetary and Space Science* 46 (1998): 1099–107.

2. Lee W. Bailey, *The Enchantments of Technology* (Champaign, IL: University of Illinois Press, 2005), 150. For a critical review of religious experiences in space, see Kendrick Oliver, *To Touch the Face of God* (Baltimore, MD: Hopkins University Press, 2013).

The Records of the Conversation

Creation Today: An Easter Debate about a Hurricane on Saturn (April 2, 2051)

1. The concept of a system tending by itself to increase its inherent order has a long history. In contemporary times, William R. Ashby introduced the term *self-organizing system*. "Principles of the Self-Organizing Dynamic System," *Journal of General Psychology* 37 (1947): 125–28. The term was taken up in system theory, e.g., Heinz von Foerster, "On Self-Organizing Systems and Their Environments," in *Self-Organizing Systems*, ed. Marshall C. Yovits and Scott Cameron (London: Pergamon Press, 1960), 31–50; in chemistry, e.g., Ilya Prigogine and Isabelle Stengers, *Order out of Chaos: Man's new Dialog with Nature* (New York: Bantam Books, 1984); in physics and biology, e.g., Erich Jantsch, *The Self-Organizing Universe* (New York: Pergamon Press, 1980).

An Eighth Day of Creation

1. New Creation: 2 Corinthians 5:17; cf. Isaiah 43:18–19, 65:17. Eighth Day: Epistle of Barnabas 15.8–9; Justinus, Dialogue 138.1. For the background, see Jürgen Moltmann, *God in Creation: A New Theology of Creation and the Spirit of God* (Minneapolis, MN: Fortress Press, 1985), 276–96. A molecular-biological textbook covers the subject from a different perspective: Horace F. Judson, *The Eighth Day of Creation. Expanded Edition* (New York: Cold Spring, 1996).
2. Persian king: Isaiah 41:1–7, 45:1–6; Ezra 1:1–2. New exodus: Isaiah 40:3–5, 42:9–10, 43:14–21, 48:6–7, 20–21.
3. Tacitus, *Histories* 5.2.4; Tibullus, *Elegies* 1.3.18.
4. Matthew 7:16–20, 12:33; Luke 6:44.
5. See, e.g., Christopher Bryan, *The Resurrection of the Messiah* (New York: Oxford University Press, 2011), 14–17, 35–41, 59–60, 159–72.
6. For participatory perceptions, see Arnold Benz, *Astrophysics and Creation: Perceiving the Universe Through Science and Participation* (New York: Herder, 2016), 63–65, 100–04.
7. Hesiod, *Works and Days* 169–73 (trans. Glenn W. Most, Loeb Classical Library, Cambridge, MA: Cambridge University Press, 2006); cf. Pindar, *Olympic Odes* 2.70–74. The myth of Cronus is reported by Hesiod, *Theogony* 137–38, 459–67. In addition, there is the tradition that Cronus was the ruler of the legendary Golden Age, where people lived like the gods and death was only sleeping. Hesiod, *Works and Days* 109–26.
8. *Peratai* (traversers), according to Hippolytus, *Refutatio* 5.16.2–3. Quotation from *Hippolytus: Refutation of All Heresies*, ed. and trans. M. David Litwa, *Writings from the Greco-Roman World*, vol. 40 (Atlanta: SBL Press, 2016), 297–99.
9. Raymond Klibansky, Fritz Saxl and Erwin Panofsky, *Saturn and Melancholy: Studies in the History of National Philosophy, Religion, and Art* (Nendeln, Liechtenstein: Kraus Reprints, 1979), 151–59, 209–14.
10. Georg Fischer, William S. Kurth, Donald A. Gurnett, et al., "A Giant Thunderstorm on Saturn," *Nature* 475 (2011): 75–77.
11. *Arithmetic complexity* is a well-defined notion in information theory, measuring the information necessary to reproduce a series of numbers. Series of the same number or repeating sequences have a small complexity. The required information is largest—and therefore also the largest arithmetic complexity—in a series of random numbers that never repeat themselves. The universe was simpler in its early phase and since then through structure formation has increased immensely in complexity.

From the Old Cosmos to the Modern Universe

1. Apocryphal Life of Adam and Eve 13–16; Qur'an, Surah 7:11–18; 15:26–43.
2. Matthew 27:45–54; apocryphal Gospel of Peter 15–22.
3. Cf. Proverbs 8:22–31; Sirach 24:1–22; Wisdom 7:22–8:1.
4. Colossians 1:15–20. For Christ as the "Intermediary of Creation," see also John 1:1–14; 1 Corinthians 8:6; Hebrews 1:2–3; further Revelation 3:14; Colossians 3:11; Ephesians 1:23; as well as Hans Weder, in *Kosmologie und Kreativität: Theologie und Naturwissenschaft im Dialog*, by Jürgen Audretsch, Hans Weder, and Markus Huppenbauer (Leipzig: Evangelische Verlagsanstalt, 1999), 60–66, 72–79.
5. *Orphic Fragment* 21a (= Ps.-Aristotle, *De mundo* 7:401a28–b7). Concerning discussion with the modern pantheism, see John Polkinghorne, *Science and Providence* (London: SPCK, 1989), 18–35.
6. Cf. Psalm 148; Daniel 3:51–90 (Greek); Revelation 5:13.
7. Revelation 22:13; cf. 1:17–18.

Creation in Different Worldviews

1. The relevance of theology for today's worldview was discussed by Georg Picht, in *Theologie—was ist das?* [Theology—what is it?], ed. Georg Picht and Enno Rudolph (Stuttgart: Kreuz, 1977), 134–37; and Adolf M. K. Müller, *Die präparierte Zeit* [The prepared time], 2nd ed. (Stuttgart: Radius, 1973).
2. Heraclitus, *frag.* 53.
3. The B ring is the densest and most prominent of the three main rings of Saturn. Science information from Britt R. Scharringhausen, Lucy Lim, Philip D. Nicholson, Keith Matthews, and Peter J. McGregor, "Ground-Based Observations of the 10 August 1995 Saturn Ring-Plane Crossing," *Icarus* 154 (2001): 287–95.

Different Perceptions in Different Perspectives

1. John Polkinghorne, *Belief in God in an Age of Science* (New Haven, CT: Yale University Press, 1998).
2. Ian G. Barbour, *Religion and Science* (San Francisco: Harper, 1997), 77–105.
3. Federico Tosi, Diego Turrini, Angioletta Coradini, and Gianrico Filacchione, "Probing the Origin of the Dark Material on Iapetus," *Monthly Notices of the Royal Astronomical Society* 403 (2010): 1113–30.

4. Yvonne J. Pendleton, Cristina M. Dalle Ore, Roger N. Clark, and Dale P. Cruikshank, *Organic Molecules on the Surfaces of Iapetus and Phoebe*, AAS 49 (2017) #210.05.

The Language of Images

1. We follow the analyses of Hans Weder, in *Kosmologie und Kreativität*, 54–60; Hans Weder, "Mythos und Metapher," in *Bibel und Mythos*, ed. Bernd Jaspert (Göttingen: Vandenhoeck, 1991), 38–73; and Georg Picht, *Kunst und Mythos* (Stuttgart: Klett, 1986), 523–30.
2. 1 John 1:5; cf. Psalm 27:1, 36:10; about Jesus Christ: John 1:5, 8:12, 9:5, 12:46; as well as the allegory of the cave by Plato, *Republic*, 516b, 517c.
3. Arnold Benz, *The Future of the Universe: Chance, Chaos, God?*, 2nd ed. (New York: Continuum, 2002), 161.
4. Carl Friedrich von Weizsäcker in a talk at an interdisciplinary meeting in Heidelberg in 1983 (recorded by Samuel Vollenweider).

Quantum Mechanics and Reality

1. The metaphor of God as the Big Observer was proposed by Jean Guitton in *Dieu et la science: Vers le métaréalisme* [God and science: Toward the meta realism], by Jean Guitton, Igor and Grichka Bogdanov (Paris: Grasset, 1991), 147ff.
2. The philosopher Bernulf Kanitscheider, for example, emphasized the importance of the interpretation in *Von der mechanistischen Welt zum kreativen Universum* [From the mechanical world to the creative universe] (Darmstadt, Germany: Wissenschaftliche Buchgesellschaft, 1993), 105–20.
3. According to modern interpretations of quantum mechanics, every observation raises a decoherent "history" of the universe. In the tree structure of all possible occurrences, one process becomes reality; e.g., Murray Gell-Mann, *The Quark and the Jaguar* (New York: Freeman, 1994), 135–65.

And Endlessly Surges the Quantum Ocean

1. Ian J. R. Aitchison, "Nothing Is Plenty," *Contemporary Physics* 26 (1985): 333–92.
2. In 1925, the American physical chemist Robert S. Mulliken discovered a small shift in the spectral lines of boron monoxide, whose energy came evidently from "nowhere." Only two years later, Werner Heisenberg, the well-known quantum physicist, could explain this energy of the vacuum with the new quantum mechanics.

3. Walter T. Grandy, Jr., *Relativistic Quantum Mechanics of Leptons and Fields* (Dordrecht, Netherlands: Kluwer, 1991), 206ff.
4. Adam G. Riess, Robert P. Kirshner, Brian P. Schmidt, et al., "BVRI Light Curves for 22 Type I A Supernovae," *Astronomical Journal* 177 (1999): 707–24; and Saul Perlmutter, Greg Aldering, Gerson Goldhaber, et al., "Measurements of Omega and Lambda from 42 High-Redshift Supernovae," *Astrophysical Journal* 517 (1999): 565–86.
5. William James, *The Varieties of Religious Experience: A Study in Human Nature* (New York: Longmans & Green, 1914), 242–43, 511–12, 523–24; William James, *A Pluralistic Universe* (Lincoln: University of Nebraska Press, 1996), 288–93.

Creation without Interruption

1. E.g., suggested by Christian Link, *Die Welt als Gleichnis* [The world as a parable], 2nd ed. (Munich: Christian Kaiser, 1982).
2. Matthew 6:25–34, Luke 12:22–31; cf. Matthew 10:29–31, Luke 12:6–7.
3. The James Webb Space Telescope (JWST) is a next generation flagship astrophysics mission. It has more than twice the diameter of the *Hubble* Telescope. Launch is planned for 2021.

Why So Uncertain?

1. E.g., Fritjof Capra, *The Tao of Physics* (Berkeley: Shambhala, 1975); Sal Restivo, *The Social Relations of Physics, Mysticism, and Mathematics: Studies in Social Structures, Interests, and Ideas* (Dordrecht, Netherlands: D. Reidel, 1983), 1–156. Cf. Kocku von Stuckrad, *The Scientification of Religion: A Historical Study of Discursive Change, 1800–2000* (Boston: de Gruyter, 2014), 87–93.
2. Aristotle, frag. 196 (= Porphyry, *Life of Pythagoras* 41); Clemens, Stromata 5.50.1.
3. *"Peratai"* in Hippolytus, Refutatio 5.14.1. Quotation from Litwa, *Writings from the Greco-Roman World*, 285–87.
4. Johann Wolfgang von Goethe, *Faust*, part 2, final chorus of *Chorus mysticus*.
5. Stanton J. Peale, "Origin and Evolution of the Natural Satellites," *Annual Reviews of Astronomy and Astrophysics* (1999): 533–602.
6. Jonathan I. Lunine, William C. Tittemore, "Origins of Outer-planet Satellites," in *Protostars and Planets III*, ed. Eugene H. Levy and Jonathan I. Lunine (Tucson: University of Arizona Press, 1993), 1149–76.

In the Beginning Was the Vacuum

1. The possibility that the universe originated out of a vacuum fluctuation has been repeatedly discussed since the 1950s. Edward P. Tryon first published this idea in "Is the Universe a Vacuum Fluctuation?" *Nature* 246 (1973): 396; Alexander Vilenkin expanded it with the tunnel effect in "Creation of the Universes from Nothing," *Physics Letters* B 117, no. 1–2 (1982): 25.

2. Alan H. Guth, *The Inflationary Universe* (Reading, MA: Perseus, 1997), 1–15.

3. Jens D. Colditz, *Kosmos als Schöpfung* [Cosmos as creation] (Regensburg: S. Roderer, 1994), 161ff.

4. 2 Maccabees 7:28; Wisdom 11:17; Hebrews 11:3; Plato, *Timaeus* 47e–52d.

5. Ilya Prigogine and Isabelle Stengers, *The End of Certainty: Time, Chaos, and the New Laws of Nature* (New York: Free Press, 1997), 179–82.

6. For example, Augustine, *Confessions* 11.:12.14ff.

7. See also the discussion between William L. Craig and Adolf Grünbaum in *Philosophia naturalis* 31 (1994): 217–49.

8. James B. Hartle and Stephen W. Hawking published another version of the basic idea of the origin out of the vacuum, which does not go beyond the zero point of time and manages without a singularity and preconditions about the vacuum in "Wave Function of the Universe," Physical Review D 28 (1983): 2960–75.

9. Genesis 1:1–5; Isaiah 45:6–7; cf. 2 Maccabees 7:28.

10. Romans 4:17; 1 Corinthians 1:28; 2 Corinthians 4:6, 4:16; Martin Luther, *The Small Catechism in The Book of Concord: The Confessions of the Evangelical Lutheran Church*, trans. by Theodore G. Tappert (Philadelphia: Fortress Press, 1987), 344-45. The relation between creation and justification is discussed in Ulrich H. J. Körtner, *Solange die Erde steht* [As long as the earth exists] (Hannover: Lutherisches Verlagshaus, 1997), 44–51.

At the River of Time

1. Plato, *Politicus* 268d–71a.

2. Further explicated in Christian Link, *Schöpfung* [Creation], vol. 2 (Gütersloh: Gütersloher Verlagshaus, 1991), 594–95.

3. The reader may refer also to the text by Knud E. Løgstrup, *Schöpfung und Vernichtung* [Creation and destruction], German translation (Tübingen: Mohr, 1990); also Gerd Theissen, *The Shadow of the Galilean: The Quest of the Historical Jesus in Narrative Form* (Minneapolis: Fortress Press, 2007), 160–61, 167.

4. Luke 24:39–40; John 20:20, 25, 27.
5. "Christ Is Risen" is a German Easter hymn from the twelfth century. The second stanza of unknown authorship got lost in the official English translation by Catherine Winkworth (1854). Translation by the authors.
6. Cf. Augustine, *De civitate Dei* 7.19; Max Pohlenz, article "Kronos," in: *Paulys Realencyclopädie der classischen Altertumswissenschaft* [Encyclopedia of antiquity], 2nd ser., vol. 11, ed. Georg Wissowa and Wilhelm Kroll (Stuttgart: J.B. Metzlersche Verlagsbuchhandlung, 1922), 1993–98.
7. On the following topic see Hesiod, *Theogony* 453–500; Macrobius, Saturnalia 1.8.7–12; Cornutus, *Nat. deor.* 6.20–7.5; John the Lydian, *Mens.* 1.1; Varro in Augustine, *De civitate Dei* 7.19; Proclus, *In remp.* 2.61.22; Cicero, *Nat. deor.* 2.64.
8. Even the Latin name *Saturnus* lets itself be associated with the insatiable devouring of time, cf. Cicero, *Nat. deor.* 2.64: "He was named Saturn because he satiated himself with years (*saturaretur annis*). The poets claim the god had repeatedly devoured his children."
9. Ovid, *Metamorphoses* 15.234–36. Quotation from Ovid, *Metamorphoses*, vol. 2, trans. Frank J. Miller, Loeb Classical Library (London: Heinemann, 1958), 381.

The Birth of Time

1. Albert Einstein, *The World as I See It* (New York: Open Road, 2011), 22.
2. The sentence is attributed to Niels Bohr.
3. Albert Einstein, Max and Hedwig Born, *Letters 1916–55* (London: Macmillan, 1971), 82 (04/29/1944), 149 (09/07/1944), 199 (10/12/1953); and also Werner Heisenberg, *Physics and Beyond* (New York: Harper and Row, 1971), 62–69.

Does God Play Dice?

1. Manfred Eigen and Ruthild Winkler, *Laws of the Game: How the Principles of Nature Govern Chance* (Princeton, NJ: Princeton Science Library, 1993).
2. Proverbs 8:30–31. For theological perspectives on the game theory, cf. Georg Picht, in *Theologie—was ist das?*, ed. Georg Picht und Enno Rudolph (Stuttgart: Kreuz Verlag, 1977), 503–8.
3. Psalm 19:1, music by Franz Joseph Haydn, *The Creation*, 1798.
4. Plato according to Plutarch, *Table Talk* 8.2.718c: "God is always doing geometry." Quotation from Plutarch, *Moralia* 9, trans. Edwin L. Minar, Francis H. Sandbach, and William C. Helmbold, Loeb Classical Library (London: Heinemann, 1961), 119.

5. Quoted from a letter from Einstein dated March 21, 1955, after learning of the death of his friend Michele Besso, as quoted in Max Jammer, *Einstein and Religion: Physics and Theology* (Princeton, NJ: Princeton University Press, 1999), 161.

6. Georg Picht, *Theorie und Meditation*, in: Georg Picht, *Hier und Jetzt* [Here and Now]. *Philosophieren nach Auschwitz und Hiroshima* [Philosophizing after Auschwitz and Hiroshima] (Stuttgart: Klett, 1980), vol. 1, 391–406.

7. Mircea Eliade, *The Myth of Eternal Return or Cosmos and History*, Bollingen Series (Princeton, NJ: Princeton University Press, 2005).

8. Matthew 13:44–46, 18:23–35, 20:1–16; Luke 16:1–8, 14:15–24.

9. Luke 12:39–46; Matthew 24:43–51 (cf. 1 Thessalonians 5:2; Revelation 3:3; 2 Peter 3:10); Luke 12:13–21, 17:26–37.

10. Genesis 6:5–7, 8:21–22; cf. 1 Samuel 15:11–35; Karl Löning and Erich Zenger, *Als Anfang schuf Gott: Biblische Schöpfungstheologien* [As a beginning God created: Biblical creation theologies] (Düsseldorf: Patmos, 1997), 167–73.

11. Exodus 33:18–22.

12. Exodus 33:18–22.

13. Carl F. von Weizsäcker, *Major Texts in Physics*, ed. Michael Drieschner (Cham: Springer, 2014), 168. See also Prigogine and Stengers, *End of Certainty*.

14. Boethius, *Consolatio philosophiae* 5.6.4.

15. Leonardo Borgarelli, Eastwood Im, William T. K. Johnson, and Luca Scialanga, "The Microwave Sensing in the Cassini Mission: The Radar," *Planetary and Space Science* 46 (1998): 1245-56.

Models—Constructs or Proxies for the Truth?

1. Attempts regarding super gravitation and common unified theories were published already in the 1970s. Great strides, but still no conclusion, could be attained in the 1980s. A representative, easily readable summery was published by Steven Weinberg, *Dreams of a Final Theory: The Scientist's Search for the Ultimate Laws of Nature* (New York: Pantheon Books, 1992).

2. Rebecca A. Harbison, Peter C. Thomas, and Philip C. Nicholson, "Rotational Modeling of Hyperion," *Celestial Mechanics and Dynamical Astronomy* 110 (2011): 1–16.

3. According to Jacques Laskar, Bureau des Longitudes [French Bureau of Longitude] (Paris), 1989.

Chaotic Prospects

1. For a survey article regarding chaos in quantum mechanics, see e.g. Eric J. Heller and Steven Tomsovic, "Postmodern Quantum Mechanics," *Physics Today* (July 1993): 38–46.
2. David Jewitt and Nader Haghighipour, "Irregular Satellites of the Planets: Products of Capture in the Early Solar System," *Annual Review of Astronomy and Astrophysics* 45 (2007): 261–95.

In the Shadow of Catastrophes

1. Review article by Paul D. Spudis, "The Moon," in *The New Solar System*, ed. J. Kelly Beatty et al. (Cambridge, MA: Cambridge University Press, 1999), 136–38.
2. Stephen J. Gould, *Wonderful Life: The Burgess Shale and the Nature of History* (New York: W. W. Norton, 1990), 13ff., 45–52.
3. Cf. Georges I. Gurdjieff, *All and Everything: Beelzebub's Tales to His Grandson*, 1st ser. (San Diego: Harcourt, 1950), 79ff., 686ff.; Arthur R. Peacocke, *Theology for a Scientific Age* (Minneapolis MN: Fortress Press: 1993), 262–64. Cf. Fred Hoyle, *The Origin of the Universe and the Origin of Religion* (Wakefield RI: Moyer Bell, 1993).
4. Martin Luther 1524, from the Latin *media vita in morte sumus*.
5. Fred Hoyle, *The Black Cloud* (London: Penguin, 2010).

Extraterrestrial Intelligences and Their Religion

1. Origen, *Commentary on John* 1.31.216–219; *De principiis* 4.4.4.
2. Joel 2:28–30; quoted in Acts 2:17–18.
3. Joel 2:30.
4. Acts 26:26 ("for this was not done in a corner"); contrary to Celsus, a second century opponent of Christians: see Origen, *Contra Celsum* 6.78.
5. Cf. Stephen J. Gould, *Time's Arrow, Time's Cycle* (Cambridge, MA: Cambridge University Press, 1990), 1–20.
6. Saying of Theodorus of Asine (according to Proclus, *In Tim.* 1.213). See also Picht and Rudolph, *Theologie—was ist das?*, 252, 512.
7. Blaise Pascal, *Pensées* (New York: E. P. Dutton, 1958), *frag.* 206.

Life—the Most Beautiful Child of the Universe

1. Referring to the German Christmas song "Lo, How a Rose E'er Blooming" from the sixteenth century. The rose is a metaphor for Jesus, born into a hostile world.

2. Richard Dawkins, *The Blind Watchmaker* (New York: Norton, 1986).

3. Joseph Ford, in *The New Physics*, ed. Paul Davies (Cambridge, MA: Cambridge University Press, 1989), 365–72; Prigogine and Stengers, *The End of Certainty*, 137–44.

4. Pascual Jordan, "Die weltanschauliche Bedeutung der modernen Physik" [The Philosophical Significance of Modern Physics], in: Hans-Peter Dürr (Publ.), *Physik und Transzendenz* [Physics and Transcendence] (Bern: Scherz, 1986), 224–226; see also Pascual Jordan, *Der Naturwissenschaftler vor der religiösen Frage* [The Natural Scientist faced with the Religious Question] (Oldenburg: Stalling, 1963), 346–356.

"And God Saw That It Was Good"? Creation Is Evaluated

1. Genesis 2:18–24, 6:5–7; cf. the Mayan book of creation: Dennis Tedlock, trans., *Popol Vuh* (New York: Touchstone, 1996), 63–69.

2. In Christian history of theology, the doctrine of God's "self-emptying" played only a marginal role before the nineteenth century. Its origin is a "kenotic" statement about Christ in Philippians 2:6–7 ("who, though he was in the form of God, did not regard equality with God as something to be exploited, but emptied himself"). The notion became important in American process theology; see John B. Cobb and David R. Griffin, *Process Theology: An Introductory Exposition* (Louisville, KY: Westminster, 1976), 41–62. The Jewish philosopher Hans Jonas interpreted the kenotic concept as an answer regarding the theodicy question. The creation of the world already appears as the consequence of a total self-emptying of God. God did not intervene at Auschwitz not because he did not want to but rather because he could not; see *Gedanken über Gott* [Thoughts about God] (Frankfurt a. M.: Suhrkamp, 1992), 33–37.

3. Joshua 10:12–14; Psalm 23:4; Mark 15:33–34.

4. Francis Crick, The Astonishing Hypothesis: *The Scientific Search for the Soul* (New York: Touchstone, 1994).

God—Mighty King or Poor Wanderer?

1. Matthew 28:20.

2. Psalm 104:29–30 (about living creatures); also Isaiah 34:4, 51:6; Psalm 102:26–28; Hebrews 1:10–12. Matthew 28:20 (about support). The topic is explicitly addressed in Walter Dietrich and Christian Link, *Die dunklen Seiten Gottes* [The dark sides of God], vol. 2 (Neukirchen, Ger.: Neukirchener, 2000).

3. Psalm 93:1.

The Universe's Hospitable Nature

1. A goal-oriented tendency would mean that the universe has the property to permit life at a certain time. This controversial postulate is also known under the name "strong anthropic principle." See also John Barrow and Frank Tipler, *The Anthropic Cosmological Principle* (Oxford: Oxford University Press, 1986), 294.

2. Cf. Peacocke, *Theology for a Scientific Age*, 226–40.

3. Dawkins, *The Blind Watchmaker*.

4. Paul Davies, *God and the New Physics* (New York: Touchstone, 1983); Paul Davies, *The Mind of God* (London: Orion, 1992); Frank J. Tipler, *The Physics of Immortality* (New York: Anchor, 1994).

5. Luke 15:17–24; cf. also John 12:32, 2 Corinthians 5:20.

Götterdämmerung of Humanity?

1. Richard E. Leakey, *The Origin of Humankind* (New York: Basic Books, 1994); Walter Scheidel, *The Great Leveler: Violence and the History of Inequality from the Stone Age to the Twenty-First Century* (Princeton: Princeton University Press, 2017). See also Carel van Schaik and Kai Michel, *The Good Book of Human Nature: An Evolutionary Reading of the Bible* (New York: Basic Books, 2016).

2. Comprehensive study was published by Georg Picht, *"Ist Humanökologie möglich?"* [Is human ecology possible?] in *Humanökologie und Frieden* [Human Ecology and Peace], ed. Constanze Eisenbart (Stuttgart: Klett, 1979), 109–22.

3. Albert Schweitzer, *"Ehrfurcht vor dem Leben"* [Reverence for life], in *Strassburger Predigten* [Strassburg Sermons] by A. Schweitzer, ed. Ulrich Neuenschwander (Munich: C. H. Beck, 1966), 135–42.

4. Cf. James, *Varieties of Religious Experience*, 355–57; Gerd Theissen, *Biblischer Glaube in evolutionärer Sicht* [Biblical faith: An evolutionary approach] (Munich: Christian Kaiser, 1984), 114–15, 146–47.

5. For everything following, see Carl F. von *Weizsäcker, Wahrnehmung der Neuzeit* [Perception of the modern era] (Munich: Hanser, 1984), 19–35 (with two texts from Georg Picht).

6. Genesis 3:5, 22.

7. Genesis 11:1–9.

8. Proverbs 16:18.

9. Of Babylon: Revelation 14:8, 18:1–24.

10. Orphism is a religious movement (initially shamanism) in ancient Greece that was organized in private cults and spread by oracle priests. Orpheus was considered its mythical founder. Regarding the myth, compare Orphic fragments 34, 35, 207, 208, 210, 214, 224, and esp. 220. Oympiodor, *Comm. in Phaed.* 1.3; Nonnus, *Dionysiaca* 6.155–75.

11. An early myth, handed down in many fragments, narrates that after achieving world rule, Zeus in the guise of a snake, copulated with his own daughter Persephone. Out of this union came Dionysus, the god of wine and ecstasy. Zeus placed him on the throne before the other gods as their young king. Driven by envy, the sinister Titans lured the childlike god with toys out of the palace, ripped him apart into seven portions, and ate him. Athena, the goddess of insight and the arts, could save his heart, which still lived. Full of rage, Zeus struck the Titans dead with his lightning. From the soot of the smoke of their bodies he formed humankind. Zeus made a "new Dionysus," as rescuer of accursed humanity, out of the heart of the killed boy.

12. Karl Kerényi, *Prometheus: Archetypal Image of Human Existence* (Princeton, NJ: Princeton University Press, 1997).

13. For Greek reference to crucifixion, see Lucian, *Prometheus* 1ff.

14. Aeschylus, *Prometheus* 907–27; Pindar, *Isthmian Odes* 8.31–35; Apollodorus, *Library* 3.169.

In the Chat Room of the Journalists (June 19)

1. Ingo Müller-Wodarg, Caitlin A. Griffith, Emmanuel Lellouch, and Thomas E. Cravens, eds., Titan: Interior, Surface, Atmosphere, and Space Environment (Cambridge, UK: Cambridge University Press, 2014).

Logbook Entries from May 19–20

1. The existence of inland seas or huge lakes on the surface of Titan was confirmed by the Cassini spacecraft. Ellen R. Stofan, Charles Elachi, Jonathan I. Lunine, et al., "The Lakes of Titan," *Nature* 445 (2007): 61–64. Waves on these lakes are generally tiny, but there are gigantic storms (Ricardo Hueso and Agustín Sánchez-Lavega, "Methane Storms on Saturn's Moon Titan," Nature 442 [2006]: 428–31) can roughen the surface (Alexander Hayes, Marco Mastrogiuseppe, Ralph Lorenz, et al., "The Depth, Composition, and Sea State of Titan's Mare," *AAS/Division for Planetary Sciences Meeting Abstracts* 46 (2014): 112.09).

Apocalypse between Physics and Theology

1. Luke 13:28; Matthew 8:12; 13:42, 50, 22:13, 24:51, 25:30; here, however, not as a result of the cold but from agony.

2. Concerning the development of the sun and the universe, cf. e.g. Arnold Benz, *The Future of the Universe: Chance, Chaos, God?*, 2nd ed. (New York: Continuum, 2002), 141ff.

3. Freeman J. Dyson addressed the future biological development in "Time Without End: Physics and Biology in an Open Universe," *Review of Modern Physics* 51 (1979): 447–60.

4. For more on the subject, see John Polkinghorne and Michael Welker, eds., *The End of the World and the Ends of God: Science and Theology on Eschatology* (Harrisburg, PA: Trinity Press, 2000).

5. Isaiah 13:10, 34:4; Ezekiel 32:7–8; Joel 3:4, 4:15; Mark 13:24 (Matthew 24:29; Luke 21:25ff.); 2 Peter 3:10; Revelation 6:12–14, 8:12, 16:8–9, 20:11, 21:1.

6. 4 Ezra 14:50.

7. Revelation 21:1–22:5; cf. Isaiah 65:17, 65:25, 66:22.

8. Luke 17:20–21.

9. Mark 13:33–35; Mark 14:34, 38; Matthew 24:37–44, 25:13 26:38, 41; Luke 12:35–40, 17:26–37, 21:36; cf. 1. Thessalonians 5:2–8; Revelation 3:3, 16:15; Acts 20:31; 1 Corinthians 16:13; Colossians 4:2; 1 Peter 5:8.

10. Philo of Byblos, *frag.* 2 (see Eusebius, *Praeparatio evangelica* 1.10.36–37). Quotation from Philo of Byblos, *The Phoenician History*, ed. Harold W. Attridge and Robert A. Oden Jr. (Washington, DC: Catholic Biblical Association, 1981), 59.

11. Hesiod, *Theogony* 729–43. Quotation from Most's translation, 63.

12. Ethiopic Apocalypse of Enoch 18:10–21:10. The various parts of the text are estimated to date from 300 BC to AD 100. Quotation from *The Old Testament Pseudepigrapha*, vol. 1, ed. James H. Charlesworth (Garden City, NY: Doubleday, 1983), 23.

13. Quotations from Odes of Solomon 42. Cf. 1. Peter 3:19–20; Matthew 12:40; Acts 2:24; Revelation 1:18; Gospel of Peter 41–42; Gospel of Nicodemus 17–27. Quotation of Solomon's Ode from *The Old Testament Pseudepigrapha*, vol. 2, ed. James H. Charlesworth (Garden City, NY: Doubleday, 1985), 771.

14. A quote from an Orphic gold tablet found in a grave in southern Italy (about 400 BC). Quotation from Alberto Bernabé and Ana Isabel Jiménez San Cristóbal, Instructions for the *Netherworld: The Orphic Gold Tablets* (Leiden: Brill, 2008), 9.

The Outcome of the Saturn Mission, or a Space Odyssey 2051

1. Heraclitus, *frag.* 60.

Index

Abraham 176
Adam 35, 239
Alchemy 190
Alexander the Great 138
Amazement 30, 98
analogies *see* metaphors, parables
angels 26, 38, 171, 186, 193
    fallen 35, 238–39
    anthropic principle 266
anthropocentrism 75, 187
apocalypse, apocalypticism 230–33
Aristotle 26, 202, 259, 261
artificial intelligence 228
asteroid 150, 160–63, 229
astrology 68
astrophysics 135, 261
Attractor, Great (astronomy) 140

Babylon 19, 44, 85
Bacon, Francis 152
Barth, Karl 113
being and becoming 83–84
big bang 107–11, 146
black hole 229
Bohr, Niels 72–75, 100
Buddhism 96, 100, 169

Cain and Abel 207
catastrophe, cosmic 40, 161, 206
causality 94, 124–26, 201

chance, randomness 13, 73, 118, 126, 163
    and God 131–37, 180
chaos 12, 21, 128, 148–54
    mythological 80–86, 113
Christ
    descent into hell/Hades 240, 245
    and God 172
    incarnation 171, 188
christology, cosmic 37–41
Clarke, Samuel 91
comet 155, 204
complexity 36, 162, 258
consciousness 51–53, 75, 228
conservation of the world *see* creation
contingency 113
Copernicus, Nicolaus 66, 151
cosmology
    antique 43–44
    antique and modern 85
    models 108–9
    religious 97
    standard model 109
    theological 44, 88, 230
cosmos
    as body of God 40, 43
    created in the image of Christ 40, 180
creatio continua see creation: conservation of the world

creation
  completion 231–32
  conservation of the world 86–87, 112–13
  creation from nothing 44, 106, 112–13
  creation of time 88, 112
  experiment 187, 193
  internal perspective of the world 51
creation, narrative biblical 112, 187
crisis 14, 162
Cronus *see* Saturn-Cronus; time
Cross 36, 40, 118, 163, 172
  price in entropy 37

Darwinists 51, 163, 201
decay 36, 118, 121, 191, 232
  see also proton decay
descent into hell *see* Christ
design 8, 197
dialogue, interdisciplinary 54
dice God playing *see* God
doxology 194

Earth
  chaotic orbit 149
  future 225
Easter 13
  cosmic dimension 19
  origin of time 119
ecology 210
Egypt 19
Einstein, Albert 93–95, 124–28, 133–36
energy conservation 80

Enoch 238–39
Entropy 27, 199
  entropy price 36, 37
Epicurus 188
Eschatology 230
Eternity 26, 139–41, 213
  of the world 40, 44
ethics 209
evil *see* theodicy
evolution 40, 47, 51, 56, 132, 161–63
  brain 228
  of humans 205–8, 227
  mutation 75, 133, 186
  selection 119, 200
experience religious *see* perception
extraterrestrial intelligence (SETI) 169

fine-tuning, cosmic 198–99
fluctuation and structure 62, 78, 82, 88, 97, 106–10
Fortune 162
freedom 91, 128, 181
future *see* time Earth sun or universe

galaxy 140, 171–76, 197
game *see* God: game of dice
genetics 190
geometry 125, 134, 263
gnosticism 27, 99, 162, 186
goal-orientation 199, 266
God
  absence 189
  accommodation 171
  attractor 202
  coming 202

Index

game of dice 126–33, 180
light 64–65
omnipotence 132, 187, 194–95
omniscience 155, 192
presence 5, 20, 31, 64, 86, 189, 241
providence 200
remorse 137
risks 164, 192
self-emptying 188
Great White Spot (hurricane) 29

Harmony 22, 40, 120, 124, 134, 139–41
Hartle, James 109–10
Hawking, Stephen 109–10
Heisenberg, Werner 72, 75, 260
hell, Hades 174, 225, 239–45
Heraclitus 47, 245
Hermeneutics 56
Hermes 1, 136, 253
Hesiod 238
History 135, 154, 176
    cosmic 140, 162
Homer 85, 254
horror vacui 86
human being
    formation 179
    creation 45, 187, 211
hymn 37–41, 177, 194
Hyperion (moon) 148–51

Iapetus (moon) 4, 56–59
image *see* parable
irreversibility *see* time
Israel 19, 230

James, William 83
Jesus (Christ) 19, 23, 38, 40–41, 47, 89, 136, 170–72, 230
Joel 175
Jordan, Pascual 181
Jupiter *see* Zeus-Jupiter

Kepler, Johannes 151
Kubrick, Stanley 248

Laniakea (galaxy supercluster) 140
Laplace, Pierre-Simon 86, 153
laws, natural 12, 26, 30, 80, 131–35
Leibniz, Gottfried Wilhelm 90–91
life formation 179
logos, cosmic 38
Lovecraft, H. P. 183
Luther, Martin 113, 164

matter 79
meditation 96, 99
metaphor 62–67
metaphysics, antique 152, 182
    impact on natural sciences 135, 147
Minkowski, Hermann 125
miracles 13, 30, 91, 137
mission, religious 169
Mithra(s), mysteries 242
model, scientific 142–46
moons
    formation 102
    name 57
    of Saturn 148–50, 156–57
Moses 140
mysticism 22, 95–98, 178, 193
myth, mythology 67, 85, 120, 211

Neoplatonism *see* Plato, Platonism
neurochemistry 190
new creation 20, 37, 46, 119, 164
new, newness 13, 20, 118, 180
   conservation of the old 137
Newton, Isaac 72, 87, 90–91
nonlocality *see* quantum theory
nothing, nothingness 12, 78, 97, 106, 241
   creation from nothing *see* creation

observation (*see also* quantum theory) 143, 146
Odysseus 202, 238, 245
order 36, 56, 85
   order out of chaos 12, 163
Origen 171
Orpheus, Orphism 174, 211, 238, 241, 245

pantheism 39, 259
parables (*see also* metaphors) 62, 90, 100, 136, 163, 178
particle, elementary 73
   virtual 78
Paul the Apostle 23, 230
Pentecost 173
perception, experience 21, 51–53, 89, 167
participatory 24, 147, 167, 248
philosophy, Greek 86, 120, 134
physicotheology 193
plan of God *see* design
Planck, Max 147
planet formation 163
Plato, Platonism 107, 134, 177

prayer 84, 177, 194
probability 72–93, 126, 197
Prometheus 211–13
proton, decay 229

quantum ocean *see* vacuum
quantum theory
   Copenhagen interpretation 75
   entanglement 95
   nonlocality 93–95
   observation 73–75, 260
   quantum veil 73
   quantum well 71–75
   uncertainty 72, 93–95, 153
quantum vacuum *see* vacuum

reductionism 190
religions 22
   dialogue 173
resurrection 19–24, 36, 119, 163
Revelation (book by John) 230
reverence for life 209
ring system (Saturn) 26, 43, 54

Saturn (planet) 7–11, 25, 139, 152
Saturn-Cronus (god) 27, 120, 238
   chronos (time) 120–21
   Saturnalia 123, 131
   star of entropy 27
   star of the Jews 20
   star of melancholia 27
   star of the philosophers 99
   in Tartarus (prison) 238
Schweitzer, Albert 209
science, natural 143, 152
   critique 52, 147

Index

dialogue with theology 54, 201
everyday experience 66, 172
suppression of reality 46, 51
self-organization 12, 86
dissipative structures 36–37
singularity 109–11
soul 189–90
space 134, 138
expansion 79–81, 140, 199, 229
states of consciousness 21–22, 83, 208, 228
subjectivity, objectivity 5, 13, 31, 33, 51, 147
suffering 36, 40, 46–47, 118
sun, future 152, 225–29
symmetry 89, 124
sympathy 31, 177

Tao *see* yin and yang
teleology 201
Thales 85
theodicy 186, 266
theory of everything 147–48
time
kairos 118, 136
chronos *see* Saturn-Cronus
cosmic 115, 119, 139
cyclical 124, 135
fourth dimension 125, 139
future 139, 151, 225–29
god of time 120
irreversibility 135
past, present, future 141
time reversal 115
Titan (moon) 3, 25, 102–3, 164–65, 214

titanism 211
Titans (gods) 27, 211–13, 267
truth 24, 51, 145, 172

uncertainty *see* quantum theory
unity of reality or world 96, 182
universe
creation *see* cosmology big bang
expansion *see* space
future 229
Uranus (god) 27, 57, 120

vacuum 77–81
expansion of space 79, 82
primeval vacuum 106–9
vision 20–24, 248

whole 29, 52, 95
wisdom (Sophia) 37, 131
world formula 30

yin and yang 100

zero-point energy 78, 81–82
Zeus-Jupiter 27, 38, 121, 211–13, 267

Also by Arnold Benz

Astrophysics and Creation
Perceiving the Universe through Science and Participation

Paperback, 144 pp., 8.25 x 5.50, 9780824522131

Renowned astrophysicist Arnold Benz draws on his research and personal experience, as well as biblical theology, to bring science and theology into a dialogue that will alter the way you see life. Benz insists that human perception reaches further than science and demonstrates this in various examples, personal, biblical, and literary. Engaging stargazers from Isaac Newton and Pierre Simon Laplace to Walt Whitman and Immanuel Kant, along with the Christian gospels, Benz brings into focus the varying perspectives forming the bases of science and theology and their relationship to one another. Benz's illuminating argument aims to show how science and faith can be compatible. An eye-opening read!

"A sparkling gem lying in a vast field of books of science and religion. Reading it is a religious experience in itself."

—**Jesse Thomas**, professor emeritus, San Diego State University; author, *The Youniverse: The Spirit of the Twenty-First Century*

"Benz … describes brilliantly the wonders revealed by modern astronomy about the origin and structure of galaxies, stars, and planets. But he also demonstrates with great clarity how other levels of reality lying in a different plane or dimension are an equally important part of human experience and a search for meaning."

—**Eric R. Priest**, FRSE FRS, renowned physicist and mathematician; professor emeritus, University of St. Andrews, UK